The Space Treaties at Crossroads

George D. Kyriakopoulos • Maria Manoli
Editors

The Space Treaties at Crossroads

Considerations de Lege Ferenda

Springer

Editors
George D. Kyriakopoulos
National and Kapodistrian University
of Athens
Faculty of Law
Athens, Greece

Maria Manoli
Institute of Air and Space Law
Faculty of Law
McGill University
Montréal, QC, Canada

ISBN 978-3-030-01478-0 ISBN 978-3-030-01479-7 (eBook)
https://doi.org/10.1007/978-3-030-01479-7

Library of Congress Control Number: 2018960836

This Springer imprint is published by the registered company Springer Nature Switzerland AG
The registered company address is: Gewerbestrasse 11, 6330 Cham, Switzerland

Foreword

A pre-Socratic Greek philosopher, Heraclitus, once proclaimed, "the only thing that is constant is change." On the other hand, in all the legal systems of the world, the principles of legal certainty and non-retroactivity of law have generally been achieved and maintained for almost all situations and relationships among the subjects of law. Thus, from a governance perspective, there is always a constant tussle in all societies, including the international community, to strike a balance between the need to change and the desire to maintain the applicable law. In order words, at the international level, efforts have often been made to ensure that certainty of behavior according to the rule-based international legal system is not threatened and that international law is reformed and changed in order to remain appropriately relevant and meet the varying needs and aspirations of the ever-changing international community. In this regard, international space law is not an exception.

International space law is a special branch of general international law and consists of a collection of binding principles and rules, incorporated in treaties and other sources of international law, that govern outer space and outer space activities. The foremost principles of international space law are embodied essentially in five space-related treaties that have been negotiated through the United Nations General Assembly during the 1960s and 1970s. Since then, the scope and nature of global space activities and the number of space actors have changed. It is, therefore, logical to reassess the efficacy of those treaties during a period which is different from the one when they were adopted.

This book addresses the most obvious, important, and timely question: whether the five UN space treaties can respond adequately to the new dimensions of and directions in space activities. This book is a collection of selected papers by several leading experts in space law and some young professionals. What makes this book an interesting reading is that it contains analysis of various issues and it presents innovative ideas from different perspectives.

In my view, this book is worth reading, especially by those who believe that the rule-based global space governance system is essential for maintaining and strengthening the common interest of all mankind in the progress of the exploration and use of outer space for peaceful purposes.

Montreal, QC, Canada

Ram S. Jakhu

Preface

Since the beginning of the space age, the relationship between Man and outer space is continuously expanding and becoming complex. Although extensive exploration of deep space has not taken place as it was expected in the 1960s, numerous space applications - telecommunications, broadcasting media, remote sensing of the Earth, meteorology, navigation and security - have direct, important and beneficial impact on daily human life on Earth.

It is apparent that the future relationship of humankind with outer space will be extensive as well as exciting. Most notably and importantly, it is well-recognized that space activities provide practical benefits to all States. Presently, 13 States have independent launch capability, over 60 countries operate their own satellites, and worldwide there are over 70 space agencies. More than 1200 satellites are currently being operated, and thousands of new satellites (especially smallsats) are seriously under development to be launched in the near future.

The current international space law regime reflects international relations of the 1960s and 1970s, and is reflective of the views, the attitudes, the interests as well as the power relations of the space-faring States of that era. This legal regime, which throughout the years has been supplemented by soft law instruments, is mainly composed of the five United Nations space treaties. New space players, mainly consisting of non-governmental entities, which were non-existent when the current international space law regime was initially created, are fast becoming the main thrust of space activities of the future. With the global space industry as an economic activity valued at US$ 320 billion annually, a growingly influential sector called the "NewSpace Industry" is now spearheading unprecedented developments in space. These developments present serious emerging challenges to the sustainability of space activities of all nations, particularly as the international space law-making process has stagnated since the last formal space law treaty was adopted in 1979.

Within the above context, this book – which is the outcome of the International Conference on New Challenges in Space Law "The Space Treaties at Crossroads: Considerations *de lege ferenda*" held in Athens, Greece, in 2015 – explores whether the five UN Space Treaties can correspond adequately to the new realities of our time and suggests models for their adjustment to future challenges. Celebrating the 50th anniversary of the Outer Space Treaty, this book constitutes a timely collection of views, arguments, and ideas on how the future of space law should be in order to contribute to the peaceful use and exploration of outer space for the betterment of humanity. This rationale, purpose and questions constitute the crux of this book and are skillfully explored by its authors.

Athens, Greece — George D. Kyriakopoulos
Montreal, QC, Canada — Maria Manoli

Acknowledgements

This book constitutes a collection of selected papers that were presented at the International Conference on New Challenges in Space Law "The Space Treaties at Crossroads: Considerations *de lege ferenda*." The conference took place in Athens, Greece on 28–29 August 2015 and was jointly organized by the *Athens Public International Law Center (Athens PIL)* of the *National and Kapodistrian University of Athens, School of Law*, the *Institute of Air and Space Law* at *McGill University*, and the *McGill Centre for Research in Air and Space Law*.

Since 2017, the Legal Subcommittee of the Committee on the Peaceful Uses of Outer Space has focused a significant part of its works on potential models for the use and exploration of outer space in the future. The purpose of this conference and book explores the same issue from an academic perspective and seeks to contribute to this productive dialogue.

Therefore, we would like to express our utmost appreciation to all the speakers of the conference that engaged in relevant academic exchanges and to all the authors of this book, whose novel ideas will be useful in updating and restructuring the current legal framework on the use and exploration of outer space.

Furthermore, we owe many thanks to the *Athens PIL* and to the *Centre for Research in Air and Space Law* for giving us the opportunity to work on this project.

As always, we are sincerely grateful to Professor Ram Jakhu for his continuous kind advice, aspiring guidance and immense knowledge, and for his support on this work.

Athens, Greece
Montreal, QC, Canada
August 2018

George D. Kyriakopoulos
Maria Manoli

Contents

Contributors

Michael Chatzipanagiotis University of Cyprus, Nicosia, Cyprus

Amalia Dimopoulou Centre National d' Etudes Spatiales, Paris, France

Steven Freeland Western Sydney University, Sydney, NSW, Australia

iCourts Centre of Excellence for International Courts, University of Copenhagen, Copenhagen, Denmark

University of Vienna, Vienna, Austria

International Institute of Space Law, Paris, France

Space Law Committee, International Law Association, London, UK

London Institute of Space Policy and Law, London, UK

European Centre for Space Law, Paris, France

Australian Centre for Space Engineering Research, Kensington, NSW, Australia

Francesco Gaspari G. Marconi University, Rome, Italy

Stratis G. Georgilas Georgilas–Hatzopoulos Law Chambers, Athens, Greece

Sarah Germann Department of International Law in the Office of the Legal Adviser, Austrian Foreign Ministry, Vienna, Austria

George D. Kyriakopoulos National and Kapodistrian University of Athens, Faculty of Law, Athens, Greece

George Leloudas Institute of International Shipping and Trade Law, Swansea University, Swansea, UK

Konstantina Liperi Department of Civil Aviation, Ministry of Transport, Communications and Works, Nicosia, Cyprus

Philip De Man Master of Air and Space Law, University of Sharjah, Sharjah, United Arab Emirates

Master of Space Studies, University of Leuven, Leuven, Belgium

Maria Manoli Institute of Air and Space Law, Faculty of Law, McGill University, Montreal, QC, Canada

Eleni-Anna Mavroeidi Attorney-at-Law, Athens Bar Association, Athens, Greece

LL.M. Air & Space Law, Leiden University, Leiden, Netherlands

Alessandra Oliva Italian Ministry of Foreign Affairs and International Cooperation, Rome, Italy

Anja Nakarada Pecujlic University of Cologne, Cologne, Germany

Gerasimos Rodotheatos European Centre for Environmental Research and Training, Panteion University, Athens, Greece

Lukas Vanhonnaeker Faculty of Law, McGill University, Montreal, QC, Canada

Chapter 1
Positive Space Law and Privatization of Outer Space: Fundamental Antinomies

George D. Kyriakopoulos

Over the past few years, an intense debate on the development of private commercial activities in outer space has commenced. This discussion presupposes the existence of relevant intentions, the attraction of significant investment, and the development of the necessary technology, in order for planned private activities in outer space, such as space tourism or the exploitation of mineral resources from celestial bodies, to enjoy a promising future.

"Antinomy" is a Greek word (*αντινομία*) that refers to "a fundamental and apparently unresolvable conflict."[1] The author of this paper believes that this word emerges in every discussion on whether legal support and regulation of private activities in outer space is feasible on the basis of existing legal norms. This happens because positive international space law does not seem to support such initiatives: The fundamental five international space treaties, adopted by the international community from 1967 to 1979, constitute state agreements based, inter alia, on the principle of non-appropriation of outer space, including the Moon and other celestial bodies (Article II of the *Treaty on Principles Governing the Activities of States in the Exploration and Use of Outer Space, including the Moon and Other Celestial Bodies* (hereafter "Outer Space Treaty"),[2] while requiring the monitoring of States (which, subsequently, take on the relevant responsibility) over any private activities in outer space (Articles VI and VII of the Outer Space Treaty, Articles II and III of

[1] "Antinomy," online: Merriam Webster http://www.merriam-webster.com/dictionary/antinomy (last accessed on 12 August 2018).

[2] *Treaty on Principles Governing the Activities of States in the Exploration and Use of Outer Space, including the Moon and Other Celestial Bodies*, adopted on 19 December 1966, opened for signature on 27 January 1967, entered into force on 10 October 1967, 610/U.N.T.S./205 [Outer Space Treaty].

G. D. Kyriakopoulos (✉)
National and Kapodistrian University of Athens, Faculty of Law, Athens, Greece
e-mail: yokygr@law.uoa.gr

G. D. Kyriakopoulos, M. Manoli (eds.), *The Space Treaties at Crossroads*,
https://doi.org/10.1007/978-3-030-01479-7_1

the *Convention on International Liability for Damage Caused by Space Objects* (hereafter "Liability Convention").[3] Consequently, the said treaties leave little room for private activities in outer space.

Furthermore, the existing space law takes the direction of a collective exploration and exploitation of outer space, through the concept of "province of all mankind" (Article I of the Outer Space Treaty) as well as it requires, at least in principle, the adoption of a collective exploitation regime of the Moon and the celestial bodies (Article 11, paragraph 5 of the *Agreement governing the Activities of States on the Moon and Other Celestial Bodies* (hereafter "Moon Agreement")).[4]

A discussion on the specific manifestations of this antinomic relationship between existing provisions of international space law and the growing desire for involvement of private interests in outer space constitutes the subject of this paper.

1.1 Public Law V. Private Activities

It is commonly known that space exploration was initially undertaken by governmental entities, being frequently a field of intense confrontation and rivalry between the two superpowers of the Cold War, the USA and the Soviet Union, which almost monopolized, for a long time, activities in outer space. Even today, the predominantly active space services (NASA, CSA, Roscosmos, JAXA, CNSA) belong to States or intergovernmental organizations, which are composed of States (ESA).

Moreover, US and Soviet space activities in the 1950s and the 1960s were of governmental nature. This "public" character of (*lato* sensu) space exploration was universal, since it also affected what was described as "use" of outer space: As it has been rightfully mentioned, "the first concrete application of space use, namely, the use of outer space for telecommunication satellites, was characterized by the public purpose."[5]

This "public purpose" is the main characteristic of positive international space law even today, since, after the Moon Agreement,[6] there was no other international space legislation of binding nature, while, at the same time, there was a trend toward

---

[3] *Convention on International Liability for Damage Caused by Space Objects*, adopted on 29 November 1971, opened for signature on 29 March 1972, entered into force on 1 September 1972, 961/U.N.T.S./187 [Liability Convention].

[4] *Agreement governing the Activities of States on the Moon and Other Celestial Bodies*, 5 December 1979, 1363 UNTS 3 (entered into force 11 July 1984) [Moon Agreement].

[5] S. Hobe, "The Impact of New Developments on International Space Law (New Actors, Commercialization, Privatization, Increase in the Number of "Space-faring Nations"), *Uniform Law Review* (2010), Vol.15 issue 3–4, 869 at 870.

[6] *Agreement Governing the Activities of States on the Moon and Other Celestial Bodies*, adopted on 5 December 1979, opened for signature on 18 December 1979, entered into force on 11 July 1984, 1363/U.N.T.S./3 (hereinafter "Moon Agreement").

the creation of soft law regulations, through the adoption of "Principles," "Guidelines," or "Codes of Conduct."[7]

Yokaris vividly explained why space law is public law:

> In the current level of evolution of space law, the relationships regulated under this branch of international law are relations between subjects of international law, States and international organizations. Therefore, contrary to the law of airspace, whose institutional framework consists of public international law conventions as well as of instruments of international uniform private law regulating activities in airspace, not only of subjects of international law but also of natural or legal persons subject to private or public law…, international space law does not yet appear under this complex form. In the field of space law, there are no such regulations that could systemically be classified under a separate category of "private space law" or "private international space law" through the institutional form of conventions of international uniform law.[8]

Given this "public" nature of positive space law, is there a place for the regulation of private activities in outer space? Article VI of the Outer Space Treaty is eloquent in providing that States shall bear international responsibility "for national activities in outer space, … *whether such activities are carried on by governmental agencies or by non-governmental entities*." Moreover, activities by nongovernmental entities in outer space "shall require authorization and continuing supervision by the appropriate State." It is thus evident that, even when they really exist, private space activities do not comprise an autonomous legal treatment, given that they are legally perceived as "annexes" to a broader activity of State character.

Or, in Yokaris' words:

> This fact is in contrast to the problem of private space activities. Indeed, private companies also participate in the space exploration and exploitation programs of States that have the necessary technology (USA, the group of countries participating in the European Space Agency, Japan), either in research, construction or in issues related to the economic exploitation of outer space – mainly in regards to satellite telecommunications. *But in all these cases, the State keeps the private activities under its control*, whether they are limited to the technical preparation of government space programs or associated with their financial exploitation. Thus, *only the State is present in the international sphere, in respect of the activities undertaken in relation to the exploration and exploitation of space*. In the context of the institutional framework composed by international conventions and agreements of the Law of Outer Space, *only the launching State, having control over all these activities, is a subject of rights and obligations and can be held internationally liable*.[9]

Similarly, Hobe shares the view for a "public" nature of the Outer Space Treaty, which is the fundamental international space law instrument, mentioning that

---

[7]…Such as the UNCOPUOS Space Debris Mitigation Guidelines (A/RES/62/217, 1 February 2008), the IADC Space Debris Mitigation Guidelines, the Principles Relating to Remote Sensing of the Earth from Outer Space (A/RES/41/65, 3 December 1986) or the EU International Code of Conduct for Outer Space Activities.

[8]A Yokaris, *International Law of Airspace and of Outer Space* (in Greek) (Greece: Ant. N. Sakkoulas, 1996), at 264–265.

[9]*Ibid*, at 265 (emphasis added).

> …the Outer Space Treaty does not fulfill [the] task [of forging an international legal order for the commercial use of outer space], given its rudimentary provisions on the freedoms of outer space which reflect the fact that outer space is the province of all mankind.[10]

Even supporters of the existence of private property rights in outer space, such as White, admit that "the paucity or outright absence of law regarding certain key subjects such as property rights, mining, salvage, liability, and dispute resolution is a disincentive to private space activities."[11]

1.2 The Non-appropriation Principle V. Property Rights in Space

Article I (2) of the Outer Space Treaty establishes the principle of freedom of exploration and use of the outer space:

> Outer space, including the Moon and other celestial bodies, shall be free for exploration and use by all States without discrimination of any kind, on a basis of equality and in accordance with international law, and shall be free access to all areas of celestial bodies.

This "freedom of exploration and use" is not limitless: It must be interpreted in the light of Article II of the same treaty, which sets out the principle "of non-appropriation":

> "Outer space, including the Moon and other celestial bodies, is not subject to national appropriation by claim of sovereignty, by means of use or occupation, or by any other means". Same principle is contained in art. 11(2) of the Moon Agreement[12]: "The Moon is not subject to national appropriation by any claim of sovereignty, by means of use or occupation, or by any other means."

The core element in Article II of the Outer Space Treaty is "national appropriation." Said wording existed in UNGA Resolution 1721(XVI) of 20 December 1961 as well as in UNGA Resolution 1962(XVIII) of 13 December 1963 (principle no 3). According to the opinions expressed during the *travaux préparatoires* of the Outer Space Treaty - and those of distinguished scholars[13] – the expression "national

---

[10] Hobe, "The Impact…," *supra* note 5, at 878.

[11] W. White, "The Legal Regime for Private Activities in Outer Space," paper presented at "Space: The Free Market Frontier" (15 March 2001), online: Space Future http://www.spacefuture.com/archive/the_legal_regime_for_private_activities_in_outer_space.shtml (last accessed on 12 August 2018).

[12] It is important that, in accordance with Art. 1 of the Moon Agreement, references to the moon in the treaty shall be understood as applicable to all celestial bodies within the solar system, other than the Earth.

[13] S Freeland and R Jakhu, "Article II," *in* S Hobe, B Schmidt-Tedd, K-U Schrogl, and G Meishan Goh Eds, *Cologne Commentary on Space Law*, Vol. 1 Outer Space Treaty (Carl Heymanns Verlag, 2009), at 50; S Hobe, "Adequacy of the Current Legal and Regulatory Framework Relating to the Extraction and Appropriation of Natural Resources," *in* Institute of Air and Space Law, McGill University, *Policy and Law Relating to Outer Space Resources: Examples of the Moon, Mars, and*

appropriation" prohibits both the exercise of sovereign rights (by States) and private appropriation (by nongovernmental entities). This conclusion is further strengthened by the clear wording of Article 11(3) of the Moon Agreement, according to which

> Neither the surface nor the subsurface of the Moon, nor any part thereof or natural resources in place, *shall become property of any State, international intergovernmental or non-governmental organization, national organization or non-governmental entity or of any natural person.*[14]

Consequently, in view of the principle of non-appropriation, it is hard to see how (State or private) property rights can be legally established in outer space. Of course, in accordance with Article VIII, paragraph 2 of the Outer Space Treaty, "ownership of objects launched into outer space, including objects landed or constructed on a celestial body, and of their component parts, is not affected by their presence in outer space or on a celestial body or by their return to the Earth." However, this article is inapplicable against resources mined on a celestial body, since they were previously covered by the non-appropriation principle and, obviously, ownership thereon is inconceivable.

1.3 Interests of Humankind V. Interests of States and Individuals

Humankind ("mankind" in treaty wording) holds a prominent place in the outer space treaties: According to Article I, paragraph 1 of the Outer Space Treaty, "The exploration and use of outer space, including the Moon and other celestial bodies, shall be carried out for the benefit and in the interests of all countries, irrespective of their degree of economic or scientific development, and shall be the province of all mankind." An analogous provision resides in Article IV, paragraph 1 of the Moon Agreement. Besides, Article V of the Outer Space Treaty considers astronauts as "envoys of mankind in outer space," while Article 11, paragraph 1 of the Moon Agreement, stipulates that "The Moon and its natural resources are the common heritage of mankind."

On the basis of these provisions, it seems that humankind per se can be validly considered as a distinct subject of international space law.[15] The acceptance of this assumption inevitably implies a right of every State to require any activity in outer

other Celestial Bodies, Workshop Proceedings, 28–30 June 2006, at 206; S Hobe, "The Legal Framework for a Lunar Base Lex Data and Lex Ferenda," *in* Gabriel Lafferranderie and Daphné Crowther, Eds, *Outlook on Space Law over the Next 30 Years: Essays Published for the 30th Anniversary of the Space Treaty* (Springer, 1997), at 138–139.

[14] Emphasis added.

[15] Yokaris, *supra* note 8, at 268–269.

space be exercised in the interest of all States, in other words in the interest of (Hu) mankind.[16]

Nevertheless, some scholars consider that a distinction must be made between the two concepts: "province of all mankind" and "common heritage of mankind." It is obvious that the supporters of private profit-making activities in space clearly adopt less holistic approaches of these concepts. Brittingham considers that the "province" concept is ambiguous and open to interpretation, as only a few States actually contribute to outer space practice, without sharing the benefits.[17] At best, "province of mankind" should give access to space resources at a fair market price, "for everyone,"[18] whereas "common heritage of mankind" should just impose a space exploration and exploitation for peaceful purposes.[19]

Tronchetti makes a concrete distinction between the two concepts:

> Under the 'province of all mankind' concept, States are free to explore and use outer space as long as they do not harm other States. Although the exploration and use of outer space shall be carried out for the benefit and in the interests of all States, the 'province of all mankind' concept does not establish any obligation to share the benefits derived from outer space activities. By contrast, under the 'Common Heritage of Mankind' concept the exploration and exploitation of a certain 'area' and its resources shall be carried out in accordance with the rules established by an international regime or authority. Successful explorers, users and exploiters, would be obliged to conform to that international regime and share the benefits derived from their exploitative activities. In such sharing of the benefits, particular regard must be paid to the developing States, regardless of their level of participation in such activities.[20]

Finally, Gabrynowicz considers that a "strategic distinction" exists between the two concepts, as "the 'province of all mankind' provision contained in the Outer Space Treaty refers to 'activities (exploration and use)'," while "the 'common heritage' provision as contained in the Moon Treaty refers to 'material objects.'"[21]

In order to reconcile opposing views on the matter, Goedhuis proposed, as a common denominator, four essential elements of the "heritage" notion that (1) the area under consideration cannot be subject to appropriation, (2) all countries must share in its management, (3) there must be an active sharing of benefits reaped from the exploitation of these resources, and (4) the area must be utilized exclusively for peaceful purposes.[22] However, one could observe that a reference to "sharing" could not be avoided.

---

[16] Ibid.

[17] Bryon C Brittingham, "Does the World Really Need New Space Law?" (2010) 12 Oregon R.I.L., 37.

[18] *Ibid*, at 39.

[19] Jeremy L Zell, "Putting a Mine on the Moon: Creating an International Authority to Regulate Mining Rights in Outer Space" (2006), 15 *Minnesota J.I.L.*, at 496.

[20] Fabio Tronchetti, *The Exploitation of Natural Resources of the Moon and Other Celestial Bodies* (the Netherlands: Nijhoff, 2009), at 44–45.

[21] Joanne Irene Gabrynowicz, "The 'province' and 'heritage' of mankind reconsidered: A new beginning," NASA Johnson Space Center, Second Conference on Lunar Bases and Space Activities of the twenty-first century, vol. 2, 1992, at 692.

[22] See Daniel Goedhuis, "Some Recent Trends in the Interpretation and Implementation of the

In any case, the cornerstone for the "province" and "heritage" concepts is the "common interest of mankind" provision: According to the Preamble of the Outer Space Treaty, there is a "common interest of all mankind" in the progress of the use of outer space, whereas such "use" "should be carried on for the benefit of all peoples, *irrespective of the degree of their economic or scientific development*." This "common interest of mankind" in outer space had already been recognized in the "historic" UNGA Resolutions 1348(XIII) of 13 December 1958, 1472(XIV) of 12 December 1959, 1721(XVI) of 20 December 1961, 1962(XVIII) of 13 December 1963, and, relatively recently, 55/122 of 27 February 2001 where a desire for "the energetic promotion" of the "fullest… exploitation of outer space for the benefit of mankind" had also been expressed.[23] Moreover, the first of the famous "principles" contained in Resolution 1962/1963 provides that "…use of outer space shall be carried on for the benefit and in the interests of all mankind." The third principle also provides that "outer space and celestial bodies are not subject to national appropriation by claim of sovereignty, by means of use or occupation, or by any other means." In particular, Resolution 1962/1963, in reaffirming previous resolutions, provides, in its Preamble, that "...Use of outer space should be carried on for the betterment of mankind and for the benefit of States irrespective of their degree of economic or scientific development."

Reference must also be made to the Declaration on International Cooperation in the Exploration and Use of Outer Space for the Benefit and in the Interest of All States, Taking into Particular Account the Needs of Developing Countries (UNGA Resolution A/RES/51/122 of 13 December 1996. According to one of the included principles:

> All States, particularly those with relevant space capabilities and with programmes for the exploration and use of outer space, should contribute to promoting and fostering international cooperation on an equitable and mutually acceptable basis. In this context, particular attention should be given to the benefit and the interests of developing countries and countries with incipient space programmes stemming from such international cooperation conducted with countries with more advanced space capabilities.

Nevertheless, as Jakhu has pointed out, the "international cooperation" mentioned in this resolution is far from taking a mandatory form, given that, according to another of the included principles, "States are free to determine all aspects of their participation in international cooperation in the exploration and use of outer space on an equitable and mutually acceptable basis."[24] It is indeed remarkable that in the case of A/RES/51/122, the focus is at the "benefit and interest of all *States*" rather than to "mankind."

In respect of the "common heritage of mankind" provision, it would be useful to look back to the origins of this concept. The relative term first appears in a proposal

---

Rules of International Space Law," (1981) 19 *Columbia J.T.L.*, at 212.

[23] See also Resolution 1348(XIII) of 13 December 1958 (above cited).

[24] Ram Jakhu, "United Nations Principles in Outer Space," in Proceedings of the United Nations/Nigeria Workshop on Space Law on "Meeting international responsibilities and addressing domestic needs" held on 21–24 November 2005, in Abuja, Nigeria, pat 28–38.

of Malta to the Secretary General of the United Nations, through its representative Arvid Pardo: By note verbale dated 18 August 1967, Malta proposed the inclusion in the agenda of the 22nd Session of the UN General Assembly the adoption of a Declaration concerning the international regulation of the seabed and the ocean floor. In a historic statement before the First Committee of the United Nations, on November 1, 1967, Pardo, inter alia, stated:

> In the light of current technological developments, however, the compromise turns out to be no compromise at all; it is clear that the sea-bed beyond the 200-metre isobath will soon be subject to exploitation. The only question is, will it be exploited under national auspices for *national* purposes, *or will it be exploited under international auspices and for the benefit of mankind*?... The sea-bed and the ocean floor are a *common heritage of mankind* and should be used and exploited for peaceful purposes and for the exclusive benefit of mankind *as a whole*.[25]

It is certainly a fact that the industrialized countries have shown no intention to share the noble aspirations of Arvid Pardo, both with respect to the ocean depths and in relation to the Moon and other celestial bodies.[26] This reluctance, derived from their technological superiority, dictated to them an approach of the "common heritage of mankind" concept, which essentially meant "all States shall have access to the outer space resources (although, for some of them, the Moon is too far)."[27] Thus, the dispute over the interpretation of the "common heritage" doctrine led to the formation of the following dilemma: equitable sharing (of benefits) or equal access (to space resources)?[28]

Through this debate, the fundamental concept is finally put forward:

Whether the outer space is either "the province of all mankind" or "the common heritage of mankind," in any case the dominant conceptual element remains the same: It is *(hu)mankind* that charges both concepts.

Humankind, as a notion, is based on consistency, not on division: It is thus a profoundly different concept from "every nation." According to Baslar, "humankind" was generally associated with the notion of "all States," while, according to another view, the term refers to "all peoples."[29] Gorove also emphasized this sense of community, when he considered that humankind describes "a collective body of peoples wherever they may be found."[30].

---

[25] Statement of Arvid Pardo, 1 November 1967, First Committee, UNGA, 22 UN GAOR, 1515th and 1516th Meeting, at 8–9 (1515th) and 2 (1516th) (emphasis by the author).

[26] Kelly M Zullo, "The Need to Clarify the Status of Property Rights in International Space Law," (2001–2002) 90 *Georgetown L.J.*, at 2424.

[27] See Ricky J. Lee, *Law and Regulation of Commercial Mining of Minerals in Outer Space* (Switzerland: Springer, 2012), at 15.

[28] Daniel A. Porras, "The 'Common Heritage' of Outer Space: Equal Benefits for most of Mankind" (2006), 37:1 *California West.I.L.J.*, at 154.

[29] Kemal Baslar, *The Concept of the Common Heritage of Mankind in International Law* (The Netherlands: Nijhoff, 1998), at 73; Leo B Malagar and Marlo Apalisok Magdosa-Malagar, "International Law of Outer Space and the Protection of Intellectual Property Rights" (1999), 17 *Boston U.I.L.J.*, at 343.

[30] Stephen Gorove, "The Concept of the Common Heritage of Mankind: A Political, Moral, and Legal Innovation?" (1972), 9 *San Diego L.R.*, at 393.

It has already been proposed that humankind constitutes a distinct (from States) subject of international space law. Although States (at least some of them) have tried to overlook the collective element contained in the "(hu)mankind" term[31] – especially since its reference to the Outer Space Treaty takes place the same year that Pardo makes his monumental speech before the First Committee (1967) – its inclusion in the existing international legal instruments adversely affects the acceptance of private business activities in space.

1.4 International V. National (Space) Law

Outer space belongs to the category of the so-called global commons, which include areas beyond national jurisdictions[32] (high seas, deep seabed, international airspace, Antarctica). This is clear, inter alia, in the combined reading of Article I paragraphs 1 and 2 of the Outer Space Treaty: "The exploration and use of outer space... shall be carried out for the benefit and in the interests of all countries... and shall be the province of all mankind. What is more, 'outer space... shall be free for exploration and use by all States without discrimination of any kind, on a basis of equality and in accordance with international law, and there shall be free access to all areas of celestial bodies." The absence of any national jurisdiction is also patent in the principle of non-appropriation, enshrined in Article II of the Outer Space Treaty, pursuant which "outer space, including the Moon and other celestial bodies, is not subject to national appropriation by claim of sovereignty, by means of use or occupation, or by any other means."

Any activity in outer space is regulated by international law. This derives from the combined reading of Articles I, III, and VI of the Outer Space Treaty. Although Article VI leaves space for "national" activities in outer space, Article III provides that "States Parties to the Treaty shall carry on activities in the exploration and use of outer space, including the Moon and other celestial bodies, *in accordance with international law, including the Charter of the United Nations,* in the interest of maintaining international peace and security and promoting international cooperation and understanding." The applicability of international law is further reinforced by Article I, where it is stipulated that "[o]uter space, including the Moon and other celestial bodies, shall be free for exploration and use by all States without discrimination of any kind, on a basis of equality *and in accordance with international law*." Further, it is important to note that Article III of the Outer Space Treaty sets out a

[31] As Mineiro observes, "the vast majority of commentary on the Outer Space Treaty that assess the concept of mankind focuses on the operative nature of the relevant treaty provisions as they relate to and among States – overlooking the possibility that mankind is a distinct holder of international legal rights" – Michael Mineiro, *Space Technology Export Controls and International Cooperation in Outer Space* (Switzerland: Springer, 2012), at 182.

[32] Surabhi Ranganathan, "Global Commons," *European Journal of International Law* (2016), 27:3,1, at 693–717.

binding obligation,[33] which was universally accepted by the main powers of the 1960s[34] during the drafting of the Outer Space Treaty.

With respect to the exploration and exploitation of outer space, international space law in force mainly comprises the Outer Space Treaty and the Moon Agreement. Up to now, 18 States have ratified the Moon Agreement which provides, in its Article 11, for an international regime of exploitation of the natural resources of the Moon, "as such exploitation is about to become feasible." For the majority of States that have not yet ratified this instrument, the international norms applicable to the exploration and exploitation of outer space are the relevant provisions of the Outer Space Treaty. In this context, particular attention must be given to the (aforementioned) articles (i.e., Articles I and II of the Outer Space Treaty).

The applicability of international law with respect to activities in outer space is, at least implicitly, recognized by national space legislations, which, in general, delimit their scope of application to matters that, although related to outer space, are closely linked with territorial sovereignty and state jurisdiction on Earth: For instance, the Belgian Space Law of 17 September 2005[35] and the Greek Law 4508/2017 mainly regulate topics such as the licensing of space activities, the transfer of activities, national registration of space objects as well as liabilities, and counterclaims and measures in the event of falling space objects.[36] What is more, the primacy of international law is almost explicitly recognized, as Articles II and IV of the Belgian Law provide that its applicability is limited, *rationae loci*, to places under the jurisdiction and control of Belgium, whereas the applicability of international norms with respect to outer space activities, enshrined in Article III of the Outer Space Treaty, is solemnly repeated[37]:

> Article 2, paragraph 1. This law covers the activities of launching, flight operations and guidance of space objects carried out by natural or legal persons in the zones placed under

[33] See Olivier Ribbelink, "Article III," *in* S Hobe, B Schmidt-Tedd and K-U Schrogl, *Cologne Commentary on Space Law,* Vol. 1, Outer Space Treaty (Carl Heymanns Verlag, 2009), at 66–67.

[34] USSR, the UK, and the USA.

[35] As revised by the Law of 1 December 2013. More information herein online: Belgian Science Policy Office http://www.belspo.be/belspo/space/belaw_en.stm (last accessed on 12 August 2018).

[36] Belgian Law of 1 December 2013, articles 13–15 (*idem*). See also Articles 3, 10, 11, 13, and 17 of the Greek Law 4508/2017 which contains similar arrangements.

[37] However, the *US Commercial Space Launch Competitiveness Act* (H.R. 2262) seems to go to a different direction, as "commercial" exploration and recovery of space resources are encouraged and promoted, whereas US citizens are explicitly entitled, by Act of State, to "possess," "own," and "sell" asteroid resources; thus the US Act explicitly confers to US citizens' property rights to resources in outer space – Commercial Space Launch Competitiveness Act, H.R.2262, 114th Congress (2015–2016) online: https://www.congress.gov/bill/114th-congress/house-bill/2262/text (last accessed on 12 August 2018). On 25 November 2015, the Act was signed by the president and became Public Law No: 114–90 – see information online: https://www.congress.gov/bill/114th-congress/house-bill/2262/all-info (last accessed on 12 August 2018), and see George D. Kyriakopoulos, "Legal Challenges Posed by the Action of Non-State Actors in Outer Space," in M Manoli and S. B. Habchi, Eds, *Conflicts in Space and the Rule of Law*, Monograph Series V (Montreal: McGill Centre of Research in Air and Space Law, 2017), at 273–290, at 283 *et seq*.

the jurisdiction or control of the Belgian State or using installations, personal or real property, owned by the Belgian State or which are under its jurisdiction or its control.

Article 2, paragraph 2. When provided for under an international agreement, this law may apply to the activities referred to under indent 1 and carried out by natural or legal persons of Belgian nationality, irrespective of the location where such activities are carried out.

...

Article 4, paragraph 3: The activities must be carried out in accordance with international law and, in particular, with the principles laid down in the Outer Space Treaty and the other treaties and agreements to which Belgium is a party.[38]

It follows that the global governance of outer space activities is forcibly of international character. The term "governance" is hereby used with the meaning of "the way that [outer space] is managed at the highest level, and the systems for doing this."[39] Given the nature of outer space as a *global commons* and taking into consideration the explicit applicability of international law in this regard (through Article I, paragraph 2 and Article III of the Outer Space Treaty), it can be validly concluded that the rules in force with respect to the governance of the celestial bodies as well as of the "empty" space [including orbits around the Earth or a(nother) celestial body] are international rules. Consequently, activities in outer space, either public or private in nature, are governed by international law.

Such an international governance of space activities, in order to be effective, requires, in the medium or long term, the establishment of an international institutional framework. Past legal experience in this respect includes the administration of international airspace by the International Civil Aviation Organization[40] (through the recognition, to its Member States, of a series of functional jurisdictions inside the so-called flight information regions) or the administration of ocean seabed by the International Seabed Authority (United Nations Convention on the Law of the Sea (hereafter "UNCLOS"), Part XI combined with the 1994 Agreement).[41] It is hereby proposed that the establishment of an international governance framework for outer space activities constitutes a sine qua non condition for an efficient, lawful, and sustainable exploitation of outer space resources, independently of the nature of the stakeholders involved.

[38] See also Articles I(2) and III(3) of the Greek Law 4508/2017 (*idem*).

[39] See "Governance" in Cambridge dictionary, online: Cambridge dictionary https://dictionary.cambridge.org/dictionary/english/governance.

[40] Based on Article 12 of the 1944 Chicago Convention on International Civil Aviation: "...Over the high seas, the rules in force shall be those established under this Convention."

[41] UN General Assembly, Convention on the Law of the Sea, 1833 UNTS 3, entry into force 16 November 1994.

1.5 Antinomies Cannot Be Resolved Through Legal Interpretation

According to the general rule of interpretation in international law, as described in Article 31, paragraph of the 1969 Vienna Convention on the Law of Treaties (hereafter "VCLT"),[42] "[a] treaty shall be interpreted in good faith in accordance with the ordinary meaning to be given to the terms of the treaty in their context and in the light of its object and purpose." Moreover, pursuant Article 32 VCLT, recourse may have supplementary means of interpretation, including the preparatory work of the treaty and the circumstances of its conclusion, in order to, among others, determine the meaning when the interpretation according to Article 31 VCLT "(a) leaves the meaning ambiguous or obscure…." It seems that the "ordinary meaning" and "the object and purpose" criteria of the general rule, as well as the "circumstances" criterion of the supplementary means rather, reinforce the notion of "(hu)mankind" as expressed above.

1.6 Conclusion

In conclusion, the antinomies exposed hereinabove constitute, in toto, a "legal discouragement" of private activities in outer space, despite the increasingly strong intention to undertake such initiatives. In the context of a space law of profoundly public and international character, which includes humankind in its entirety as one of its subjects and dictates the non-appropriation of resources in outer space, no sufficient ground for an effective regulation of commercial activities is currently in place. This *lacuna* cannot be redressed by way of interpretation, as the relevant rules of the 1969 Vienna Convention are not able to overcome the intrinsic characteristics of the existing international legal framework.

This is not to say that commercial activities in outer space, of a private nature, should be discouraged. On the contrary, the trend shows that such activities are about to expand. For instance, States, such as the USA, tend to place the famous "use of outer space" concept within a business context. If the above proves to be true, the international space law in force might prove insufficient to address the challenge.

Moreover, one should not forget that other important issues for an effective and secure commercial exploitation of outer space – such as the protection of the space environment or the creation of a space traffic management system, which is still in an embryonic condition – will also require, in the near future, the intervention of state authorities through the undertaking of relevant international action. Therefore, the existing antinomies between the international outer space legal framework (still)

[42] Vienna Convention on the Law of Treaties, 1155 UNTS 331, entry into force 27 January 1980.

in force and the (private) activities that are underway require a substantial legal update of the existing international norms (e.g., through improvement, renovation, and/or change), especially given their inadequacy in terms of legal security, as soft law principles and guidelines do not suffice to bring the necessary stability and certainty.

Chapter 2
Interpreting the UN Space Treaties as the Basis for a Sustainable Regime of Space Resource Exploitation

Philip De Man

2.1 Introduction

Long considered only from a theoretical perspective, a spate of initiatives revolving around the exploration and utilization of the Moon and asteroids, both public and private, have recently reintroduced the relatively long-standing United Nations (UN) treaties to the pressing issue of natural resource appropriation.[1] In direct response to the budding development of an American space mining industry,[2] the United States adopted the Commercial Space Launch Competitiveness Act in November 2015.[3] This Act was the first legal instrument to explicitly grant property

[1] The most high-profile of these initiatives are the Planetary Resources (www.planetaryresources.com), Deep Space industries (deepspaceindustries.com) and Bigelow Aerospace (http://bigelowaerospace.com). The volatility of the budding space mining market means that it is uncertain which of these companies will be the first to embark on a space mining mission, and when that will be. Recent shake-ups in the industry have already led to internal reorganization of Planetary Resources and an indefinite postponement of their first planned mission.

[2] The frontrunners in the race to space resources listed in the above footnote are private companies with headquarters in the United States: Planetary Resources (Washington), Deep Space Industries (California) and Bigelow Aerospace (Nevada). However, some of these companies have branched out into other countries, notably Luxembourg and Mexico, while new competitors are increasingly setting up shop in other jurisdictions as well, such as iSpace in Japan.

[3] H.R.2262—114th Congress (2015–2016), introduced by Rep. Kevin McCarthy.

P. De Man (✉)
Master of Air and Space Law, University of Sharjah, Sharjah, United Arab Emirates

Master of Space Studies, University of Leuven, Leuven, Belgium
e-mail: pman@sharjah.ac.ae; philip.deman@kuleuven.be

G. D. Kyriakopoulos, M. Manoli (eds.), *The Space Treaties at Crossroads*,
https://doi.org/10.1007/978-3-030-01479-7_2

rights to private enterprises over resources extracted from asteroids and other celestial bodies. In particular, the law provides that

> A United States citizen engaged in commercial recovery of an asteroid resource or a space resource [...] shall be entitled to any asteroid resource or space resource obtained, including to possess, own, transport, use, and sell the asteroid resource or space resource obtained in accordance with applicable law, including the international obligations of the United States.[4]

The reference to possession and ownership of space resources obtained in accordance with the international obligations of the United States may prove challenging in light of the current state of international space law. Indeed, as noted during congressional hearings, resource extraction represents 'a very volatile and contentious issue at the international level',[5] as also recognized by the United States in the most recent session of the Legal Subcommittee (LSC) of the UN Committee on the Peaceful Uses of Outer Space (COPUOS).[6] In anticipation, the US Act was careful to refer, on several occasions, to the international obligations of the United States on the issue of commercial recovery and exploration of space resources.[7] However, the legal implication of these references is unclear should the interpretation prevail that space resource appropriation, as such, is currently unlawful under international law. The US Congress itself is keenly aware of this possibility, by noting that the 2015 Act rests on a specific reading of international space law:

> It is the sense of Congress that by the enactment of this Act, the United States does not thereby assert sovereignty or sovereign or exclusive rights or jurisdiction over, or the ownership of, any celestial body.[8]

Though included in the corpus of the Commercial Space Launch Competitiveness Act, this disclaimer, as a sense of Congress provision, merely expresses the opinion of the US legislator and hence has no force of law.[9] Nevertheless, sense of Congress provisions do generally indicate a shift in foreign policy of the United States, and they are hence closely watched by other States.[10] In space resource matters, the move by the United States has indeed inspired other countries to move forward with

---

[4] Sec. 51,303. Asteroid resource and space resource rights.

[5] Remarks by Prof. Joanne Gabrynowicz: *see* J. Foust, 'Hearing Raises Questions About Asteroid Mining Bill', SpaceNews, 10 September 2014, available at http://spacenews.com/41825hearing-raises-questions-about-asteroid-mining-bill/, retrieved on 9 January 2018. *See* also M. Listner, 'Asteroid Resource Rights Will Require White House Support', SpaceNews, 22 September 2014, available at http://spacenews.com/41954letter-asteroid-resource-rights-will-require-white-house-support/, retrieved on 9 January 2018.

[6] UNCOPUOS, Report of the Legal Subcommittee on its 56th session, held in Vienna from 27 March to 7 April 2017, UN Doc. A/AC.105/1122 of 18 April 2017, para. 236.

[7] See Sec. 51302 (a) and (b) and Sec. 51303 of the Act. The same specific language referring to the international obligations of the United States does not figure in the other, more expansive, titles of the Commercial Space Launch Competitiveness Act.

[8] Sec. 403. Disclaimer of extraterritorial sovereignty.

[9] C.M. Davis, 'Sense of resolutions and provisions', 16 May 2016, Congressional Research Service 7-5700, available at https://fas.org/sgp/crs/misc/98-825.pdf, retrieved on 9 January 2018.

[10] Ibid.

their own agendas and indirectly resulted in the adoption of the Law on the Exploration and Use of Space Resources by Luxembourg. Although clearly inspired by the US Act, the Luxembourg initiative goes further in boldly declaring space resources to be simply 'capable of appropriation', leaving out any reference in this key provision to the country's international obligations.[11] Other countries, including the United Arab Emirates and Japan, are actively considering adopting national legislation containing provisions on space resource exploitation as well.

Even if no national activities of space mining have so far been authorized under any of the above national legal frameworks, the mere fact of their adoption is construed by some as contravening the explicit prohibition of national appropriation of outer space, including the Moon and other celestial bodies, contained in Art. II of the Treaty on Principles Governing the Activities of States in the Exploration and Use of Outer Space, Including the Moon and Other Celestial Bodies (Outer Space Treaty [OST]),[12] and Art. 11 of the Agreement Governing the Activities of States on the Moon and Other Celestial Bodies (Moon Agreement [MA]).[13] If the former provision, due to its general formulation, is unclear in its application to the natural resources of the Moon and near-Earth asteroids, the latter clearly proscribes the appropriation of natural resources in place on celestial bodies. Both provisions, however, are marred by interpretative difficulties and a lack of follow-up in terms of exploitation of material resources in space. In direct response to the developments at the national level, the UNCOPUOS LSC has, since 2016, invited discussions under a new agenda item on 'general exchange of views on potential legal models for activities in exploration, exploitation and utilization of space resources'.

Those unsympathetic to the application of existing space norms to the exploitation of resources from celestial bodies might argue that the OST is irrelevant in this respect, for it fails to explicitly address exploitation activities. To the extent that it allows the exploitation of celestial bodies, the Moon Agreement might nevertheless be deemed equally irrelevant, given its inapplicability to those countries that have adopted national laws on space resource exploitation, none of whom are parties to the Agreement. Alternatively, it could be argued under the so-called *rebus* sic *stantibus* doctrine that the circumstances of the conclusion of the space treaty frame-

---

[11] Such reference was included in a previous version of the bill, which provided that '[s]pace resources are capable of being appropriated under international law'. However, the reference to international law was dropped in a rather contentious move after the Luxembourg Council of State remarked that it would be premature and not conducive to legal certainty to assume that the matter of space resource appropriation had been settled under international law. See Ph. De Man, 'Luxembourg law on space resources rests on contentious relationship with international law', October 2017, *The Space Review*, available at http://thespacereview.com/article/3355/1, retrieved on 9 January 2018.

[12] *Treaty on Principles Governing the Activities of States in the Exploration and Use of Outer Space, Including the Moon and Other Celestial Bodies*, 27 January 1967, 610 UNTS 205, 18 UST 2410, TIAS No 6347, 6 ILM 386 (entered into force on 10 October 1967) [Outer Space Treaty].

[13] *Agreement governing the Activities of States on the Moon and Other Celestial Bodies*, 5 December 1979, 1363

UNTS 3 (entered into force 11 July 1984) [Moon Agreement].

work have changed so fundamentally since its initial formulation as to preclude its application to activities that were not realistically considered by its drafters.

Indeed, it might be argued that the negotiations of the UN space law framework did not primarily concern themselves with the regulation of the commercial exploitation of asteroids by private entrepreneurs. However, this position must be reconciled with the reality that, a decade before and ever since the adoption of the Outer Space Treaty in 1967 and the development of its principles over the course of 12 years and four conventions, spacefaring activities have always depended on the use of natural resources. Though primarily regulated in the conventions of the International Telecommunication Union (ITU) as limited natural resources, the UN space treaties confirm – via their history and wording – the legality and indeed necessity of the exploitation of orbits and frequencies for all scientific and commercial spacefaring endeavours.[14] If the applicability of these treaties to the use of intangible resources is hence uncontested, their relevance for the exploitation of material space resources, such as hydrogen, helium-3 and plutonium-238, remains unsettled.

In this context, it is often thought that the material nature and physical destruction of natural resources indistinguishable from their origin as parts of non-appropriable celestial bodies pose insurmountable problems, in light of the national appropriation proscription codified in the Outer Space Treaty and the Moon Agreement. Suggested solutions vary from a drastic reinterpretation of the fundamental provisions of the basic UN space treaties scarcely supported by their actual language (e.g. carving out resources from the celestial body notion in the indiscriminate phrasing of Article II OST) to proposals for an altogether new treaty dealing specifically with the issue of exploiting the mineral resources from celestial bodies, as suggested by the Moon Agreement. However, most of these proposals appear either internally inconsistent or politically unfeasible in the current climate.[15]

Awaiting further developments in this field, the present article therefore aims to clarify that it is possible to interpret the existing UN framework of international space law in a manner that aligns the planned activities of commercial exploitation of outer space with the language of the OST and MA, without subverting the inten-

---

[14] The treaties were negotiated and adopted at a time when the space age had already become a reality through the launching and orbiting of a number of satellites. Indeed, none of the provisions in the UN space treaties can be construed as excluding the freedom to use orbits and frequencies around Earth. Rather, by establishing an enabling regime of spacefaring activities reliant on orbital movement and radiofrequency communications, the treaties accept and presume the necessity of exploitation of these resources. This is confirmed by the fact that a number of provisions in the UN space treaties explicitly rely on orbital criteria for determining their scope of application (e.g. Art. II of the *Convention on Registration of Objects Launched into Outer Space*, 6 June 1975, 28 UST 695, 1023 UNTS 15 (entered into force 15 September 1976) [Registration Convention]), and the fact that the lowest possible orbit is often posited as a working criterion for delimiting the airspace boundary and defining the geographical scope of application of space law.

[15] For an in-depth discussion of these proposals and their shortcomings, see Ph. De Man, *Exclusive Use in an Inclusive Environment: The Meaning of the Non-Appropriation Principle for Space Resource Exploitation*, Berlin, Springer, 2016, Chapter 3.

tion of the drafters. Indeed, such interpretation may even be necessary in order to arrive at a sustainable development of space resource exploitation, by preserving the balance between the rights of some States to use resources and the interests of all other States that may lack such capacities. These findings are buttressed by an analysis of the ITU regulations on orbital usage, which still represent the most complex and detailed regime of space resource exploitation to date, and are in line with the fundamental principles dictated by the UN space treaties.

2.2 Treaty Interpretation

Though the product of a specific political climate is dictated by tensions between two cold war superpowers, the existing space treaties are also the result of a careful balancing act between space *haves* and space *have nots* that aimed to reconcile the need to facilitate spacefaring activities with their sustainability. Any conclusion that the existing space treaties are not appropriate to regulate novel activities must also recognize the need to recalibrate this precarious balance. Before questioning the continuing relevance of the UN space treaties for activities of exploitation and consumption of celestial resources, we must therefore set out the ground rules for interpreting the international framework of space law conventions, so as to determine their internal consistency, flexibility and robustness for incorporating future developments. Only if we find that the main principles of international space law, as interpreted consistently across several treaties, fail to provide a workable framework for the exploitation of the resources from celestial bodies, should it be concluded that the UN space treaties need to be adapted or perhaps even discarded.

The 1967 Outer Space Treaty, drafted by the UNCOPUOS LSC, is generally considered the cardinal instrument of international space law that lays down the fundamental principles on the exploration and use by States of outer space, including the Moon and other celestial bodies. Over the course of 12 years, the general principles of the Outer Space Treaty were elaborated in four binding instruments on rescue and return, liability, registration and the exploration and use of celestial bodies, drafted in relatively rapid succession by the UNCOPUOS LSC.[16] In light of this remarkable sequence, the UN Office for Outer Space Affairs has described the UN regime on space law as a coherent system, founded on the OST and concretized in four subsequent treaties, which 'may be said to deal specifically with certain concepts included in the 1967 Treaty'.[17] While this qualification confirms the historical

[16] These are Outer Space Treaty; *Agreement on the Rescue of Astronauts, the Return of Astronauts and the Return of Objects Launched Into Outer Space*, 22 April 1968, 672 UNTS 119, 19 UST 7570, TIAS No 6599, 7 ILM 151 (entered into force 3 December 1968) [Rescue Agreement]; *Convention on International Liability for Damage Caused by Space Objects*, 29 March 1972, 961 UNTS 187, 24 UST 2389, 10 ILM 965 (1971) (entered into force 1 September 1972) [Liability Convention]; Registration Convention; and Moon Agreement.

[17] UNOOSA, United Nations treaties and principles on outer space, UN Doc. ST/SPACE/11/Rev.2 of 2008, p. vi. The most recent collections of the UN space treaties and resolutions omit the intro-

and conceptual primacy of the Outer Space Treaty, it also underscores the particular relevance of the subsequent treaties for illuminating the general rules of the 1967 convention.[18]

This is true in particular for the Moon Agreement, whose preamble not only 'recalls' the Outer Space Treaty and its other three predecessors but explicitly 'takes into account' 'the need to define and develop the provisions of these international instruments in relation to the Moon and other celestial bodies, having regard to further progress in the exploration and use of outer space'.[19] The close nature of the relationship between the provisions of the Moon Agreement and the Outer Space Treaty is also underscored by the preparatory works of the former, during which it was noted that 'the discussion of the Outer Space Committee [sic] confirmed the understanding that the moon treaty [sic] in no way derogates from or limits the provisions of the 1967 Outer Space Treaty'.[20]

The internal coherence of the international space law regime as a network of conventions gradually crystallizing the general principles of one founding instrument renders the post-1967 conventions particularly valuable as elements of external context for interpreting the Outer Space Treaty, as set out in Article 31 (3) VCLT. The clear statement of object and purpose of the MA as intended to define and develop the principles of the OST may well qualify it, under Article 31 (3), sub a VCLT, as an 'agreement regarding the interpretation' of the Outer Space Treaty or 'the application of its provisions', many of which are identically phrased, to the sphere of the Moon and other celestial bodies in our solar system.

In this respect, Schwarzenberger defines 'interpretation' as 'the process of establishing the legal character and effects of a consensus achieved between parties', while 'application' specifically refers to 'the process of determining the consequences of such an interpretation in a concrete case'.[21] Linderfalk considers every action taken by an applier in accordance with the provision of a given treaty to

ductory part that contains this characterization: *see* UNOOSA, *United Nations Treaties and Principles on Outer Space*, UN Doc. ST/SPACE/61/Rev.1, *s.d.*

[18] J.I. Gabrynowicz, 'Space Law: Its Cold War Origins and Challenges in the Era of Globalization', 37 *Suffolk U. L. Rev.* 2004, p. 1042: '[t]ogether, the treaties form an interrelated legal framework that governs international space activities, complete with laws and dispute resolution mechanisms'.

[19] PP 7 MA.

[20] UNGA, Preparation of an international convention on principles governing the use by states of artificial Earth satellites for direct television broadcasting: report of the Special Political Committee, UN Doc. A/34/664 of 22 November 1979, cited in K.U. Pritzsche, *Natürliche Ressourcen im Weltraum – das Recht ihrer wirtschaftlichen Nutzung*, Frankfurt am Main: Lang, 1989, p. 95. Christol confirms that '[t]he Moon Treaty [sic], like all of the other space environment treaties following the 1967 Principles Treaty, fortifies and extends certain critical provisions of the latter': C.Q. Christol, 'The Common Heritage of Mankind Provision in the 1979 Agreement Governing the Activities of States on the Moon and Other Celestial Bodies', 14 *Int'l Law.* 1980, p. 480. *See* further E. Brooks, 'Dangers from Asteroids and Comets: Relevance of International Law and the Space Treaties', 40 *Proc. Coll. L. Outer Space* 1997, p. 255.

[21] G. Schwarzenberger, 'Myths and Realities of Treaty Interpretation: Articles 27–29 of the Vienna Draft Convention on the Law of Treaties', 9 *Va. J. Int'l. L.* 1968, p. 7.

constitute an application thereof.[22] This includes cases, in particular, where 'the provisions of a treaty are the cause for concluding a new international agreement or the cause for the way the new agreement is drafted'.[23]

The Moon Agreement clearly fits this description if we consider it as concretizing the principles of the Outer Space Treaty, in particular Article II, to the exploitation of the resources of celestial bodies. However, Article 31 (3), sub a VCLT, requires that an interpretative agreement be concluded 'between the parties' of the interpreted treaty. Although the MA was negotiated by the same body as the OST, the clear discrepancy between the number of States that ultimately ratified both instruments severely limits the former convention's interpretative relevance for the parties to the latter.[24]

The systemic nature of the relationship between the main UN space law treaties nonetheless requires that it be adequately reflected in the interpretative process for some of their murkier provisions and concepts. The legal basis for such an exercise can be found, if not in Article 31 (3), sub a VCLT, then possibly in sub b of the same provision. This clause refers to subsequent practice in the application of a particular convention as an interpretative tool for its provisions. Though difficult to qualify, such practice is often deemed the preferred option for including the external context of a treaty in its interpretation, as it reintroduces the intention of the parties by way of its emanation in their ensuing conduct.[25]As noted by Brownlie, the notion of practice in international treaty law can encompass, among others, 'a pattern of treaties in the same form'.[26] Sorel and Boré Eveno confirm that the notion 'subsequent practice' in Article 31 (3), sub b VCLT, can include unilateral acts as well as 'subsequent (non-interpretative) agreements'.[27] In support of this reading, they refer to case law of the ICJ that amply draws on practice of States Parties to the interpreted treaty in the form of successive conventions, in addition to the practice of bodies of relevant international organizations.[28] Other international and regional courts, such

---

[22] U. Linderfalk, *On the Interpretation of Treaties: The Modern International Law as Expressed in the 1969 Vienna Convention on the Law of Treaties*, Dordrecht: Springer, 2007, p. 164–165

[23] Ibid., p. 167.

[24] While the Outer Space Treaty has been ratified by 107 States, the Moon Agreement only has 18 ratifications (as of April 2018). For an up-to-date status of the UN space treaties, see http://www.unoosa.org/oosa/en/ourwork/spacelaw/treaties/status/index.html, retrieved on 5 April 2018.

[25] G. Distefano, 'La pratique subséquente des états parties à un traité', 40 *Ann. Fr. Dr. Int'l* 1994, p. 46; J.-M. Sorel and V. Boré Eveno, 'Article 31 1969 Vienna Convention', in O. Corten and P. Klein (eds.), Y. Le Bouthillier, 'Article 32 1969 Vienna Convention', in O. Corten and P. Klein (eds.), *The Vienna Conventions on the Law of Treaties: A Commentary*, Oxford: Oxford University Press, 2011, p. 826.

[26] I. Brownlie, *Principles of Public International Law*, Oxford: Oxford University Press, 2008, p. 6. The latest version of the principles, no longer authored by Brownlie himself, changes the long-standing terminology of the cited phrase to 'an extensive pattern of treaties in the same terms', without further explanation as to its origins or justification: J.R. Crawford and I. Brownlie, *Brownlie's Principles of Public International Law*, Oxford: Oxford University Press, 2012, p. 24.

[27] J.-M. Sorel and V. Boré Eveno, *supra* footnote 18, p. 826.

[28] Ibid.

as the International Tribunal for the Law of the Seas, too, have relied on subsequent agreements to confirm a particular reading of a divisive treaty provision.[29]

The narrow bond between the five UN space law conventions, as characterized by their relatively rapid succession and shared subject matter, their promulgation by the same legal body and their copious links with each other through express references and verbatim reproduction of key provisions, shows that they constitute a 'pattern of treaties in the same form' as intended by Brownlie. At the very least, case law and legal literature on the topic suggest that the four UNCOPUOS treaties elaborating on the general principles of the Outer Space Treaty should qualify as a form of international practice in the application of the latter convention. The provisions of the Moon Agreement retain particular relevance in this respect, given the substantial overlap between its provisions and those of the OST. In addition, the broad interpretation of the subsequent practice notion in Article 31 (3), sub b VCLT, strongly suggests that international instruments that lack the features of the self-referential UN space law framework but contain detailed rules on related subjects, such as the ITU Constitution, Convention and Administrative Regulations, should also be taken into account in clarifying the provisions of the 1967 principles treaty, on the same level as the text and context of the latter convention.

Finally, and failing their qualification, under sub a or b of Article 31 (3) VCLT, as either interpretative agreements or subsequent practice in the application of the Outer Space Treaty, the space law conventions adopted by the UN in follow-up to the Outer Space Treaty unquestionably constitute particularly relevant rules of international law for the interpretation of the latter, in accordance with sub c of the provision in the Vienna Convention. The chronological succession of the five treaties, their internal coherence and their largely overlapping subject matter demand, at the very least, that they be taken into account in the process of illuminating the basic space law tenets, pursuant to Article 31 (3), sub c VCLT, prior to and instead of resorting to instruments of more distant and unrelated branches of international law, such as the law of the high seas. The same can be argued for the constitutive treaties of and the binding rules promulgated by the International Telecommunication Union, the conformity of whose detailed regulations with the fundamental tenets of UN space law provides fertile ground for comparative readings.

Article 31 (3) VCLT puts into perspective the general requirement that treaties be interpreted in the light of the general rules of international law in force at the time of their conclusion, so as to provide flexibility in taking into account the evolution of international law between the conclusion of the relevant treaty and the time of the dispute over its interpretation.[30] Indeed, while it may be necessary to have recourse to the circumstances surrounding the conclusion of a particular treaty, the temporal element in the interpretation of treaties is not exclusively historical, for one should also consider the developments that have taken place since its adoption. In the case concerning Legal Consequences for States of the Continued Presence of South

[29] Ibid.

[30] I. Sinclair, *The Vienna Convention on the Law of Treaties*, Manchester: Manchester University Press, p. 138–140.

Africa in Namibia,[31] for example, the ICJ emphasized the need for intertemporal interpretation, by underscoring that, although it is bound by 'the primary necessity of interpreting an instrument in accordance with the intentions of the parties at the time of its conclusion', international instruments should also be interpreted and applied 'within the framework of the entire legal system prevailing at the time of the interpretation'.[32] Further, the interpretative effort should also take into account 'the present-day state of scientific knowledge'.[33] Though the friction between an interpretation guided by the intention of the parties and the legal framework at the time of interpretation should be resolved on a case-to-case basis, Sinclair offers a sensible rule of thumb by arguing for precedence of the latter when interpreting 'certain terms in a treaty which are by their very nature expressed in such general terms as to lend themselves to an evolutionary interpretation'.[34]

The field of international space law eminently lends itself to evolutive interpretation. Its codification in advance of practical developments presupposes a level of generality demanding subsequent concretization in separate instruments dealing with specific spacefaring activities, which, in turn, shed light on the content and implications of the general rules codified in the basic treaty. Moreover, it may well be that the historical conditions at the time of concluding the main space treaties have, paradoxically, facilitated an interpretation in line with the latest technological advances in the field of spacefaring. Indeed, the uncertain outcome of the so-called space race during the Cold War, often cited as one of, if not the most important historical factor in shaping the development of space law, inspired the creation of a legal regime in advance of any manned space activity on the Moon that would secure it against any potential claims of territorial sovereignty by the first to finish the race. If the historical elements of space law have thus resulted in a regime that is lamented, often rightly so, for its generality and lack of precision, these very characteristics have also made it possible to absorb the technological developments in the field of spacefaring.

In interpreting the continuing relevance of the principles codified in the OST, we should hence also consider whether or not the relevant provisions have been retained in more recent agreements, such as the MA and the regulations of the ITU. Taking into account the VCLT provisions on subsequent agreements relating to the same subject matter, it is clear that the OST principles, even if not expressly tailored to the commercial exploitation of celestial bodies by private actors, apply to this activity if they remain substantially unaltered in later treaties that do explic-

[31] ICJ, Legal Consequences for States of the Continued Presence of South Africa in Namibia (South West Africa) notwithstanding Security Council Resolution 276 (1970), Advisory Opinion, *I.C.J. Reports* 1971, p. 16.

[32] *Ibid, para.* 53.

[33] ICJ, Kasikili/Sedudu Island (Botswana v. Namibia), Judgment, *I.C.J. Reports* 1999, p. 1060.

[34] I. Sinclair, *supra* footnote 23, p. 140. An evolutive interpretation of treaties is also advocated in, for example, ICJ Legal Consequences for States of the Continued Presence of South Africa in Namibia (South West Africa) notwithstanding Security Council Resolution 276 (1970), p. 16, para. 53; ICJ, Aegean Sea Continental Shelf (Greece v. Turkey), Judgment, *I.C.J. Reports* 1978, p. 3, para. 80.

itly address such exploitation. This warrants an interpretation of the general space law principles in line with the more specific legal rules they have prompted, as codified in, primarily, the Moon Agreement, but also non-UN conventions, such as the ITU regulations.

2.3 The Moon Agreement

The above section on treaty interpretation already argued for the relevance of the Moon Agreement as a convention consistent with the Outer Space Treaty that further clarifies the content of the latter's provisions for activities exploring and using celestial bodies. The substantial overlap between the provisions of the Moon Agreement and the 1967 principles treaty has even inspired denouncements of the former as a superfluous duplication of the latter, simply adapted to a smaller scope.[35] More significant is a joint statement issued by the States Parties to the Moon Agreement, which notes that those provisions that are unique to the Agreement are of particular importance, for they 'clarify or complement principles contained in the other outer space treaties that are applicable to the Moon and other celestial bodies'.[36] Considering the absence of formal requirements for interpretative agreements, there is no reason why this joint statement on the benefits of the MA should not be considered a 'subsequent agreement between the parties regarding the interpretation' of the MA, which, in turn, confirms the substantive consistency between the provisions of the latter treaty and the OST.[37] There thus appears little reason to interpret the MA as anything but the concretization of the principles outlined in the OST to the legal environment of celestial bodies, as emphasized by the preamble of the 1979 Agreement, which is revelatory of the goal and purpose of the convention.[38]

[35] *See* in particular the scathing assessment in S.E. Doyle, 'Using extraterrestrial resources under the Moon Agreement of 1979', 26 *J. Space L.* 1998, p. 111–128, reiterated in S.E. Doyle, 'Issues of sovereignty and private property', in M. Benkö and W. Kroll (eds.), *Luft- und Weltraumrecht im 21. Jahrhundert: liber amicorum Karl-Heinz Böckstiegel*, Cologne: Carl Heymanns, 2001, p. 313–325. *See* also the critical assessment of the various draft treaties for a Moon Agreement in E. Galloway, 'The future of international space cooperation in treatymaking', 14 *Proc. Coll. L. Outer Space* 1971, p. 244.

[36] UNCOPUOS LSC, Joint statement on the benefits of adherence to the Agreement governing the activities of States on the Moon and other celestial bodies by States Parties to the Agreement, UN Doc. A/AC.105/C.2/L.272 of 3 April 2008, p. 3–4.

[37] Art. 31 (3), sub a VCLT. *See supra*.

[38] C.Q. Christol, 'The common heritage of mankind provision in the 1979 Agreement governing the activities of States on the Moon and other celestial bodies', 14 *Int'l Law.* 1980, p. 429 and 480; H.L. Van Traa-Engelman, 'Clearness Regarding Property Rights on the Moon and Other Celestial Bodies', 39*Proc. Coll. L. Outer Space* 1996, p. 42; V. Kopal, 'United Nations and the Progressive Development of International Space Law', 1997 *Finnish Yb. Int'l L.* 1997, p. 18; R.B. Bilder, 'A Legal Regime for the Mining of Helium-3 on the Moon: U.S. Policy Options', 33 *Fordham Int'l L.J.* 2009, p. 269; and R.J. Lee, *Law and Regulation of Commercial Mining of Minerals in*

This finding has particular importance for the provisions of the Moon Agreement that deal with the appropriation and exploitation of the natural resources of celestial bodies. Contained in Article 11 MA, these provisions explicitly confirm, in unambiguous language, the application of the non-appropriation principle to exploitation activities undertaken by private actors on the Moon and other celestial bodies. As such, after noting that both the Moon and its natural resources should be considered the common heritage of mankind (para. 1), Article 11 MA verbatim reiterates Article II OST (though understandably limited in its scope to celestial bodies) (para. 2) and explains that the non-appropriation principle applies to the surface, subsurface and any part of celestial bodies or their natural resources in place, whether claimed by States, (inter)national organizations or non-governmental entities and natural persons (para. 3).

Taking into account the systemic interpretation of the UN space treaties, the specific language and structure of Article 11 MA has two important implications. First, it clarifies that the non-appropriation principle applies to both areas and natural resources, be it of outer space as such or the celestial bodies therein. Second, it confirms our presumption that the OST, even if it does not regulate the exploitation of celestial bodies as such, applies to and allows for this activity, despite the explicit prohibition of appropriation. The identical scope in terms of physical coverage of the outer space environment ('outer space, including the Moon and other celestial bodies') as provided in Articles I and II OST already implies that, whatever their relevance for the natural resources of celestial bodies and orbits around Earth may be, it must be the same for both provisions. In the absence of textual indications to the contrary, natural resources must either be covered by Articles I *and* II or by neither. The latter option sounds manifestly absurd, since it would imply that natural resources can be appropriated but not used. If resources are covered by both Articles I and II OST, however, we must conclude that they can be used though not appropriated.

Based on the above analysis, it can be stated that the identical scope of Articles I and II OST requires distinguishing between the activities of use and appropriation of natural resources, in order for these provisions to be workable in practice. It follows that, though their alleged synonymy is often taken for granted, exploitation should be clearly distinguished from appropriation.[39] Disregard for this most obvious observations when applied to material resources of celestial bodies lies at the heart of the difficulties encountered in trying to reconcile the practice of

Outer Space, Berlin: Springer, 2012, p. 186

[39] *See*, for example, H.G. Darwin, 'The Outer Space Treaty', 42 *Brit. Yb. Int'l L.* 1967, p. 282–283: 'many types of 'use' or 'exploitation' [...] are inconceivable without appropriation of some degree at least of the materials taken. Thus, mineral deposits can hardly be explored [sic*!*] without appropriation'. *See* further E. Asan, 'Law and Peace for the Celestial Bodies', 5 *Proc. Coll. L. Outer Space* 1962, p. 8; E. Husby, 'Sovereignty and Property Rights in Outer Space', 3 *J. Int'l L. & Prac.* 1994, p. 365; T. Gangale, *The Development of Outer Space: Sovereignty and Property Rights in International Space Law*, Santa Barbara: Praeger, 2009, p. 42; and S.M. Williams, 'The Growing Momentum of Satellite Broadcasting and the Geostationary Orbit', 26 *Proc. Coll. L. Outer Space* 1983, p. 47.

resource exploitation with the text of the UN space treaties, as it conceives of Article I OST as contrasting rather than dovetailing with Article II OST. The distinction is clearly undisputed, however, in the case of immaterial natural resources, which have been exploited since the beginning of the space age, yet whose subjection to the non-appropriation principle has, if anything, been affirmed ever stronger (*see infra*, Sect. 2.4).

The disjunctive interpretation of the concepts of exploitation and appropriation is supported by the text of the Moon Agreement as the only legally binding international convention explicitly dealing with both types of activities relating to natural resources in outer space. In this respect, it can hardly be a coincidence that the drafters of the 1979 Agreement adhered to a conscientious terminological distinction between the 'exploitation' of resources on the Moon and other celestial bodies in the preamble and the sections of Article 11 projecting the need for a future international regime, on the one hand (paragraphs 5 through 8 of the provision), and the verbatim reiteration and clarification of the non-appropriation principle of the Outer Space Treaty in the most absolute of terms in Article 11 (2) and (3) MA, on the other hand. Hence, Article 11 MA, through its language as well as its structure, confirms the unlawfulness of the appropriation of natural resources from celestial bodies while allowing their exploitation.

To be sure, the language of Article 11 (3) MA, noting that natural resources 'in place' cannot be appropriated, is rather infelicitous, for it appears to allow resources to be appropriated once removed from their place. This fails to take into account, however, that equally authoritative versions of the Moon Agreement in other languages do not necessarily support this implied reading of Article 11 (3) MA.[40] In any event, the ambiguity inherent in this provision should not be mistaken all too readily for an affirmation that resources can be imbued with property rights when excavated. For it is unclear why the drafters of the MA would opt for an implied permission of resource appropriation through *a contrario* formulations instead of an explicit affirmation in clear terms.

More fundamentally, such an a contrario interpretation 'seems to permit somewhat questionable outcomes, as it does not sound tenable that if property rights are not allowed in space it would nevertheless be possible to claim that the space resources are legally obtainable once removed'.[41] The sudden transformation of parts of celestial bodies from non-appropriable territory to appropriable resources, merely through the act of possession/excavation, erodes the non-appropriation principle by officially sanctioning the establishment of exclusive titles in resources through the very act of physical exclusion of other users. It remains, however,

---

[40] For example, the French version of the provision extends the application of the non-appropriation provision to the natural resources that 'can be found on' celestial bodies (*'les ressources naturelles qui s'y trouvent'*).

[41] L. Viikari, *From Manganese Nodules to Lunar Regolith: A Comparative Legal Study of the Utilization of Natural Resources in the Deep Seabed and Outer Space*, Rovaniemi: University of Lapland, 2002, p. 111. *See* also K.N. Rao, 'Editorial comment: common heritage of mankind and the Moon Treaty', 21 *Indian J. Int'l L.* 1981, p. 277.

wholly unclear how these resources can be distinguished from their celestial bodies of origin. Viikari thus rightly observes that '[i]n such a case, resources to which an entity never had any rights while in outer space could become the possession [sic] of that entity merely as a result of the fact that they have been extracted - *regardless of the legality of that extraction*'.[42]

The MA, or any other space treaty for that matter, does not regulate the act of *extraction* as a legally significant activity on par with exploitation or use, nor does it confer any property rights to the excavating party.[43] Extraction merely signifies physical possession, which does not carry with it any prerogatives of exclusion, as the ITU regime makes perfectly clear (*see infra*, Sect. 2.4). Hence, it is likely that the 'in place' criterion in Article 11 (3) MA should be read as an (imperfect) expression of the legal requirement of human activity to render these resources distinguishable from their environment (i.e. exploitation), rather than as an indication of the physical location of natural resources on celestial bodies. To be sure, the ordinary meaning of the term 'in place' appears to contradict this interpretation. However, it is a general rule of interpretation of treaty provisions that the ordinary meaning of terms should be abandoned if it renders the significance of a provision manifestly absurd or unreasonable.[44] For the aforementioned reasons, it is submitted that a locational interpretation of the in place criterion in Article 11 (3) MA could hardly contribute to a sensible reading of this provision.

Therefore, the 'in place' criterion should be interpreted as referring to the act of exploitation, which in turn determines the qualification of natural resources. If we accept this interpretation, however – and logic dictates that we should – the automatic removal of resources from the scope of the non-appropriation principle through their actual exploitation becomes unworkable in practice. This is evident when we consider the regulation of orbits. If we accept that there is no a priori distinction between orbits as areas and orbits as resources, then any argument for their subjection to or removal from the scope of Article II OST based on the fact that they cannot be materially separated from their environment would be wholly arbitrary. A better understanding of this exploitation/appropriation dichotomy intrinsic in international space law requires us to leave the UN framework and closely consider the ITU regulations of orbital usage.

---

[42] Ibid., footnote 303 (emphasis added).

[43] Art. 6 MA on samples notoriously stops short of conferring property rights on the collecting State: *see* S. Hobe and F. Tronchetti, '[Moon Agreement:] Article 6 (scientific investigations/samples/minerals)', in S. Hobe, B. Schmidt-Tedd and K.-U. Schrogl (eds.), *Cologne Commentary on Space Law. Volume II: Rescue Agreement, Liability Convention, Registration Convention, Moon Agreement*, Cologne: Carl Heymanns, 2013, no. 92.

[44] Art. 32 VCLT.

2.4 Orbits and the International Telecommunication Union

If the key to resolving the natural resource conundrum lies in parsing the relationship between the activities of exploitation and appropriation, rather than in comparing superficial physical features of resources, the detailed rules of the International Telecommunication Union on the placement of satellites in orbit provide an excellent case study, as they have been tried, tested and adapted to the realities of spacefaring for over half a century. Moreover, they are binding on the entire international community, as the ITU counts no fewer than 193 States among its members.[45]

Firstly, when considering orbits, the issue of potentially disparate legal statuses is removed from the picture, for their exploitation concerns the use of a resource that is at the same time classified as a spatial phenomenon of outer space sensu stricto and as a part of celestial bodies, depending on the applicable legal text.[46] Secondly, the example of orbits epitomizes the functional nature of natural resources, for they depend on the activity of placing an artificial object in space for their existence. Finally, and most importantly, the current ITU regime on orbital usage results from a well-documented tug-of-war between two conflicting approaches to resource appropriation, in turn revealing two possible, inverse relations between exclusion and exploitation, yet both ultimately guided by a recognition of the need to sustainably regulate the use of a limited resource.

To illustrate the intricacies of the interplay between these conflicting approaches and what they reveal about the principles guiding the exploitation of space resources in general, we may refer to a statement by Robert Jones, who, during his tenure as head of the ITU Radiocommunication Sector, noted that

> [t]he single most important issue [for the ITU] is the reservation of capacity without actual use […]. Eliminating or minimizing the opportunity to acquire uncommitted resources could help alleviate the current orbital congestion. […] Recent experience indicates that unrealistically long operational lifetimes are notified, leading to almost permanent occupation of orbital positions.[47]

The cited passage first addresses the reservation of orbital capacity without actual use, also referred to as the 'paper satellite' problem,[48] or as slot 'warehousing',[49]

[45] See the list of Member States on the official website of the ITU at https://www.itu.int/online/mm/scripts/gensel8, retrieved on 5 April 2018.

[46] Art. 1 (2) MA provides that '[f]or the purposes of this Agreement reference to the Moon shall include orbits around or other trajectories to or around it'.

[47] [Note], 'ITU tries to solve new problems without interfering with national sovereignty', *Satellite Wk.* 29 May 1995.

[48] Director Radiocommunication Bureau, Report on Resolution 18 of the Plenipotentiary Conference Kyoto, 1995, p. 3. 'Paper satellites' are defined as 'satellite networks in coordination or recorded in the Master Register that are not in operation and will never be brought into use'*: see* D.M. Leive, Rapporteur Group SC-4 report to the Special committee on regulatory/procedural matters devoted to Resolution 18 (Kyoto, 1994), ITU Doc. SC-RG4/54 of 25 November 1996, p. 3 and 11.

[49] J.C. Thompson, 'Space for Rent: The International Telecommunications Union, Space law, and

which is generally deemed unlawful. At the same time, the statement also highlights the problems that may arise from prolonged instances of actual use of orbital positions by those States having the capacity to launch satellites into orbit around Earth, for they may result in an 'almost permanent occupation of orbital positions', precluding States that currently lack launching capabilities from exercising their freedom to use these slots at a future point in time.

The statement and, by extension, the legal issues that arise with respect to the practices described therein reveal the existence of two apparently contradictory approaches to exclusion in the use of natural resources in space: one that construes prolonged use as amounting to appropriation and one that derides exclusion as unlawful when not accompanied by actual use. The perceived unlawfulness of two diametrically opposed practices, which nevertheless originate in the same legal regime, highlights the need to clarify the limits of the permissible types of use (and non-use) of orbital slots. We must therefore ascertain the limits of lawful exploitation of orbits as outlined in the ITU regime.

Article 44 (2) ITU CS construes radio frequencies and all associated orbits, including the geostationary satellite orbit, as limited natural resources. Their centralized regulation is therefore warranted, with a view to guaranteeing equitable access for all, in particular developing countries, while facilitating the efficient use of orbits by those States that have spacefaring capabilities. Considering that the ITU's role in space telecommunication consists of regulating the management and use of orbits and frequencies, and that its constitutive convention categorizes them as natural resources, it is uncontested that the free use of space resources is considered lawful under the ITU regime. Like Art. I OST, such free use is not unrestricted, but rather it is subject to some limitations, which mainly aim to preserve the freedom of all States to use scarce resources despite disparate capabilities, in a sustainable manner.

The main limits to the free exploitation of natural resources in the UN treaty regime are laid down in Articles I (2) and IX OST. These provisions require that outer space, including celestial bodies, be used by all States without discrimination or harmful interference, on a basis of equality. Article 44 (2) ITU CS concretizes this requirement by noting that States must use orbital resources on a rational, efficient and economic basis, so as to guarantee equitable access thereto. To be sure, the ITU treaties do not explicitly reiterate the non-appropriation principle of Art. II OST or Art. 11 (2) MA. Nevertheless, the ITU system governing the management of frequencies and associated orbital positions is revelatory, not only of the proscription of property rights over natural resources in space but also of the specific relationship between the activities of appropriation and exploitation so as to make possible, in practice, the use of space resources, without violating Art. II OST.

In practice, the rational and efficient use of limited natural resources in space is realized under the ITU regime by requiring all administrations to complete the

---

Orbit/Spectrum Leasing', 62 *J. Air L. & Com.* 1996, p. 279–331; S. Ospina, 'Regulation of space resources', in M. Benkö and K.-U. Schrogl (eds.), *Space Law: Current Problems and Perspectives for Future Regulation*, Utrecht: Eleven International Publishing, 2005, p. 263.

registration procedure outlined in Articles 9 and 11 of the ITU Radio Regulations.[50] Protection from harmful interference by other competing users of orbits and frequencies is tentatively gained upon finalization of the registration procedure with the ITU and publication in the Master International Frequency Register (MIFR). This international protection is subject to a final check, however: if a satellite system is not brought into use (i.e. regular operation) within the notified period, or if this use is discontinued, the international protection will lapse, and the reserved resources will become available for actual use by other players.[51] This important qualification to the maligned 'first-come, first-served' system of the ITU supports the contention made earlier with respect to Art. 11 (3) MA that international space law does not confer property rights through mere physical possession; otherwise, satellites would retain protection from interference even in the absence of their actual and regular operation.

In recent years, a string of financial and administrative due diligence measures has been added to the general ITU procedure to make sure that the submission notices correspond to an actual and continuous use of the resources identified by the administration's planned or existing satellite system. These measures include the payment of a registration fee and submitting a registration notice in advance that identifies a number of key characteristics of the proposed satellite system several years before it will be brought into use.[52] The initiatives aim to ensure, in part, the sustainable exploitation of space resources by requiring the actual use of resources that have been claimed by a particular administration, in order to avoid limited resources going to waste. As such, these measures constitute a clear refutation of paper satellites and label the reservation of orbital capacity without subsequent use unlawful.

On the other hand, as long as an administration actually and continually uses a satellite to exploit the economic value of a particular orbital position – in combination with a given frequency band – there is little in the ITU regulations that would confer a legally enforceable right on the part of other administrations to put an end to the registered exploitation of said natural resources. To be sure, there are ITU resolutions proclaiming (1) that the registration of a satellite system in the MIFR does not grant permanent rights over the resources identified in the notice and (2) that any priority in protection from harmful interference obtained through registra-

---

[50] For lack of space, the present article does not cover the various plans adopted by the ITU that deviate from the general registration procedure. Suffice it to say that they do not fundamentally alter the assessment of the procedure in Arts. 9 and 11 ITU RR or their link with the principles in Arts. I and II OST. For more on the ITU plans and their relevance for the non-appropriation principle, *see* Ph. De Man, 'Rights over areas vs resources in outer space: what's the use of orbital slots?', 38 *J. Space L.* 2012, p. 39–150.

[51] Nos. 11.44 through 11.49 ITU RR.

[52] *See*, for example, Resolution 49 (Rev. WRC-12) on administrative due diligence applicable to some satellite radiocommunication services, Resolution 80 (Rev. WRC-07) on due diligence in applying the principles embodied in the Constitution and Resolution 81 (WRC-2000) on the evaluation of the administrative due diligence procedure for satellite networks.

tion will be reassessed if certain basic characteristics of the registered satellite are altered.[53] In practice, however, these resolutions cannot – nor do they wish to – limit a priori the time an administration may exploit a valuable orbital position for its own exclusive use. They merely confirm that the protected exclusive use of orbital resources is dependent on the continued operation of a previously registered satellite boasting certain characteristics.

It follows that, to the extent that the non-appropriation principle of the UN space treaties is confirmed in the context of the ITU, it renders clear that unlawful appropriation will only arise if a State claims certain resources in space for its exclusive use, yet ultimately does not end up using these resources. On the other hand, permanent, though actual, use of limited resources in space is, as such, perfectly lawful, though it may be limited both in time and quantity in order to accommodate the interests of others. This is not because the ITU regime tolerates property rights over natural resources – it doesn't – but because the legality of such monopolistic use of orbits may only raise issues from the perspective of, in UN treaty terms, Article I OST, rather than Article II OST. Exclusive exploitation is only lawful when justified by actual and continued use; conversely, it is only when exclusion occurs on the basis of mere claims or physical possession not accompanied by actual exploitation that the reservation of resources is revelatory of property rights.

Not only does this approach dovetail with both components of the objective of *sustainable use*, it also corresponds to the general interpretation of appropriation in property law theory. The notion of property rights is elucidated nowhere with more clarity than in the writings of Penner.[54] Recognized as 'one of the best accounts one can find of the kinds of rights that property – as legally understood – involves',[55] Penner's approach to the idea of ownership is founded on the basic contention that property rights are typified by the specific interaction between the related rights of exclusion and use as determinants of the social relations between the owner and non-owners, including prospective users. In general, relations between a user and others with respect to the thing that is used may take on any of three following forms, with differing degrees of exclusive impact:

> [t]he right to use something so long as no one else was using it or wanted to use it is equivalent to having no right of exclusion whatsoever. The right to use something so long as one got there first is a right to exclude others while one is using something, and the right to use whenever one wants amounts to a right to exclude others whenever one decides to use something.[56]

---

[53] Resolution 2 (Rev. WRC-03) on equitable use, by all countries, with equal rights, of the geostationary satellite and other satellite orbits and of frequency bands for space radiocommunication services and Resolution 4 (Rev. WRC-03) relating to the period of validity of frequency assignments to space stations using the geostationary satellite and other satellite orbits.

[54] In particular, J.E. Penner, *The Idea of Property in Law*, Oxford: Oxford University Press, 1997

[55] L.S. Underkuffler, *The Idea of Property: Its Meaning and Power*, Oxford: Oxford University Press, 2003, p. 32.

[56] J.E. Penner, *supra* footnote 45, p. 70.

Of these, only the latter relationship accurately describes property, for it is only when the user can exclude others *at will*, depending on his/her discretionary exercise of the right to use, that (s)he is verily endowed with property rights. The discretionary element of the *right to* exclude non-owners implies that the owner is not bound by strict considerations to allow use by others whenever such use is possible, because the owner is not using the thing owned himself. Conversely, if the authority to exclude others from using a thing is merely incidental to the actual use thereof, there is no property, for the *exclusion does not derive from any right to exclude at all* but is merely a factual consequence of the actual use of an object.

The ITU regime, buttressed by the above understanding of appropriation in general property law, shows that exploitation of limited natural resources in outer space is possible without appropriation, even if the exclusion of others may in practice appear to be permanent. For the exclusion of administrations other than the registered user only exists by virtue – and for the duration – of the actual and continued use of the orbital resources by the latter. The ITU regime also shows, by confirming the gist of the non-appropriation principle and refusing to sanction exclusion that is not corroborated by use, that the non-appropriation principle of Article II OST extends to natural resources. More importantly, at least for our purposes, it does so in the context of a binding treaty system that is not only internally consistent with the UN space law framework but has been constantly updated to the changing spacefaring needs by the entire international community of 193 States.

2.5 Conclusion

Even if the adoption of an international framework on space mining will unlikely be witnessed in the near future, this paper attempted to argue that the impending exploitation of celestial bodies by private actors for commercial reasons does not necessarily require a drastic reinterpretation or modification of the existing UN space law framework to render such activities workable. The language of the Outer Space Treaty is formulated with such flexibility as to cover most foreseeable spacefaring activities, regardless of their commercial purpose or private nature. The Moon Agreement, structurally consistent with the OST, confirms the basic legality of the activity of exploitation in its consideration of the natural resources of the Moon and other celestial bodies. At the same time, Article 11 (2) and (3) MA confirm the stringent application of the non-appropriation principle to natural resources and for private actors.

Rather than posing a theoretical incompatibility, the strict distinction made in the text of the Moon Agreement between unlawful appropriation and the perfectly legal exploitation of celestial bodies should suffice to dispel the alleged convention that consumption of material resources necessarily entails or requires the creation of property rights. As demonstrated by the detailed, up-to-date regulations of the ITU, the legality of exclusive use of a strategically valuable limited natural space resource does not depend on its duration but solely on the actuality of the exploitation. There is no reason why the same rationale should not guide the exploitation of celestial bodies, especially since the fundamental provisions in Articles I and II OST do not distinguish between outer space and celestial bodies, nor between their classification as areas or resources.

Crucially, the experience of the ITU has demonstrated the continued validity and relevance of the balancing act underlying the basic principles of the Outer Space Treaty for the exploitation of natural resources in outer space, namely, the reconciliation of the interests of those that can use space resources and those that due to technological or financial constraints may not yet be able to do so. While the aim of the ITU regime to guarantee equitable access to orbital resources by all may not be directly inspired by concerns for sustainability, the denunciation of claims for future use that are not backed up by actual necessity clearly contributes to this aim as well. For, if the existing regime of the Outer Space Treaty only allows for the legally enforceable exclusion of others in the case and for the duration of actual use by one, a proper understanding of property rights has shown us that the only thing that would be gained by allowing the appropriation of space resources would be a licence to waste. It stands to reason that any attempt to alter the balance created by the principles of the Outer Space Treaty, in particular Articles I and II, would risk undermining this most basic of starting points for a sustainable regime of space mining.

Chapter 3
The Effectiveness and Applicability of the Moon Agreement in the Twenty-First Century: Will There Be a Future?

Eleni-Anna Mavroeidi

3.1 Introduction

In the 1960s, the decision to land humans on the Moon was mostly regarded as strategic political means between the two blocs in the context of the Cold War.[1] However, the interest in the Moon has been renewed since the beginning of the twenty-first century and can now be explained by factors different from the international and political prestige of space-faring nations. Scientific reasons and the will to extract and use new natural resources,[2] supported by the progress of technology, have caused human curiosity to thrive toward the exploration of the Moon during the first decades of the twenty-first century, followed by the exploration of other celestial bodies. For instance, a flight to Mars is already anticipated,[3] as, since 2007, many unmanned lunar missions have been launched by several nations.[4] Of course,

[1] Antonella Bini, *The Moon Agreement: Its effectiveness in the 21st century,* ESPI 14 Perspectives, at 1.

[2] Large amounts of aluminum, anchorite, iron, and titanium have been discovered; helium 3 that can be used as an alternative to hydrocarbon and fossil fuels; water and ice that can support manned stations on the lunar surface; solar power, orbiting solar power satellite (SPS).; "[T]he Moon, planets and asteroids have been said to contain "aluminum, calcium, carbon, chromium, gold, hydrogen, iridium, iron, magnesium, manganese, nickel, nitrogen, oxygen, platinum, silicon, titanium and water." Heidi Keefe, *Making the Final Frontier Feasible: A Critical Look at the Current Body of Outer Space Law*, 11 Santa Clara Computer & High Tech L J 345, 362 (1995).

[3] Ian Sample, *Mars one mission: a one-way trip to the red planet in 2024,* The Guardian, 9 February 2015.

[4] 21st Century Moon Missions: United States of America: "Constellation Program" of 2004: (LRO) 18 June 2009. Its return to the Moon is estimated by 2020. The scope of this program is to establish an extended human presence on the Moon and use the Moon as a basis for more ambitious projects in the future.

E.-A. Mavroeidi (✉)
Attorney-at-Law, Athens Bar Association, Athens, Greece

LL.M. Air & Space Law, Leiden University, Leiden, Netherlands

G. D. Kyriakopoulos, M. Manoli (eds.), *The Space Treaties at Crossroads*,
https://doi.org/10.1007/978-3-030-01479-7_3

such missions require their governance by the rule of law. In this context, the Moon Agreement plays a central role in providing the legal framework according to which economic or exploration missions should take place as the Agreement is focused on the regulation of human activity on celestial bodies. Therefore, it can be predicted that natural space resources exploitation will constitute the main focus of future commercial and private interests in the Moon. Hence, the relevance of the question of the status of the Moon from a legal perspective is more topical than ever before.

3.2 The Adoption of the Moon Agreement

3.2.1 Negotiation Procedure

Taking into consideration the aforementioned missions and the need for regulation of space activities driven by humans, the maxim *nécessité fait loi* ("necessity makes law") could be brought in the fore. Indeed, the United Nations, through the Committee on the Peaceful Uses of Outer Space (COPUOS) and its Legal Subcommittee (LSC), recognized early enough the necessity to regulate such activities and took the lead in the development and codification of a relevant legal regime. Specifically, it all started with the Declaration of Legal Principles for Governing Space Activities[5] and continued with the adoption of the five UN space treaties.[6] These treaties provide, among other, humanitarian obligations for mutual assistance, the imposition of third-party liability upon launching states, and the creation

Japan: "Selene Mission" of 2007: The mission's aim was to analyze the Moon's origin and its physical composition.

India: "Chandrayan 1 Mission" of 2008: The Indian mission was aiming to investigate the various minerals and chemical substances on the Moon.

China: "Lunar Exploration Program CLEP" also known as "Chang'e Program": Three lunar robotic missions, Chang'e 1 (2007–2009), Chang'e 2 (2010), and Chang'e 3 (2013). Chang'e 5's sample return mission is scheduled for 2017. CLEP is also paving the way for the first manned mission by 2025–2030.

[5] General Assembly Resolution 1721 (XIV): In 1963 the General Assembly finally adopted the Declaration of Legal Principles Governing Activities of states in the Exploration and Use of Outer Space, as proposed by the Committee on the Peaceful Uses of Outer Space ("COPUOS"). Most of the principles embodied in the Declaration are reflected in the Outer Space Treaty.

[6] General Assembly Resolution 1962 (XVIII): Treaty on Principles Governing the Activities of States in the Exploration and Use of Outer Space, Including the Moon and Other Celestial Bodies (Outer Space Treaty, 1967), Agreement on the Rescue of Astronauts, the Return of Astronauts and the Return of Objects Launched into Outer Space (Rescue Agreement, 1968), Convention on International Liability for Damage Caused by Space Objects (Liability Convention, 1972), Convention on Registration of Objects Launched into Outer Space (Registration Convention, 1975), Agreement Governing the Activities of States on the Moon and Other Celestial Bodies (Moon Agreement, 1979).

For further information, see: http://www.unoosa.org/oosa/en/SpaceLaw/treaties.html.

of a United Nations register of space objects.[7] The Moon Agreement is the fourth child of the Treaty on Principles Governing the Activities of States in the Exploration and Use of Outer Space, Including the Moon and Other Celestial Bodies, in order to promote the exploration and use of outer space, carried out for the benefit and in the interests of all countries.[8]

The history of the Moon Agreement began in 1970[9] when Argentina presented a "Draft Agreement on the Principles of Governing Activities in the Use of the Natural Resources of the Moon and Other Celestial Bodies"[10] to the LSC. In May 1971, the Soviet Union proposed the addition of "The Preparation of an International Treaty Concerning the Moon" and submitted, some days later, its own draft Moon Agreement.[11] On 24 November 1971, the General Assembly of the United Nations (UNGA) adopted Resolution 2279 (XXVI) in which the Soviet Draft was taken into account and requested the COPUOS and the LSC to consider the issue of a Moon Agreement as a matter of priority. This question was deliberated and developed by the LSC from 1972 to 1979. The UNGA in Resolution 34/68 of December 5, 1979, adopted the Agreement and on December 18, 1979, it opened to States for signature.[12] However, the Agreement did not come into force until June 1984 when the fifth instrument of ratification was deposited by Austria. Up to January 2018, 18 states (which surprisingly are not space-faring powers) are parties to the Treaty, seven of which have ratified the Agreement, while the rest acceded.[13] Four additional states have signed but not ratified the treaty.[14] The United States, the Russian Federation (former Soviet Union), and the People's Republic of China have not signed, acceded, nor ratified the Moon Agreement, which has led to its consideration as a "failure" from the standpoint of international law.[15] Some scholars have reasoned that it is a text of "little legal value" and is "obviously unacceptable to the international community."[16]

---

[7] Article IX, Outer Space Treaty, *supra* note 12, on mutual assistance; Articles II–IV, Liability Convention, *supra* note 12 on third-party liability; Article II para 1, Registration Convention, *supra* note 12 on the registration of space objects.

[8] *Treaty on Principles Governing the Activities of States in the Exploration and Use of Outer Space, Including the Moon and Other Celestial Bodies*, 27 January 1967, 610 UNTS 205, 18 UST 2410, TIAS No 6347, 6 ILM 386 (entered into force on October, 10 1967) [Outer Space Treaty], at Annex.

[9] For a more elaborate analysis of the negotiations of the Moon Agreement and its *travaux préparatoires,* see H. W. Bashor Jr., The Moon Agreement Paradox (2004).

[10] UN Doc. A/AC. 105/85.

[11] UN Doc. A/C1/L.568, Nov. 5, 1971.

[12] G. Zhukov and Y. Kolosov, International Space Law, Novosti Press Agency, Moscow (1984) at 173; Bin Cheng, Studies in International Space Law, Oxford (1997) at 246.

[13] United Nations, *Treaty Series*, vol. 1363, *supra* note 6.

[14] Ibid.

[15] Michael Listner, *The Moon Agreement: failed international law or waiting in the shadows?* The Space Review, October 2011.

[16] E. Galloway, *Guidelines for the Review of Outer Space Treaties,* Proceedings of the 41st Colloquium on the Law of Outer Space (1998) at 2; G. H. Reynolds, *International Space Law: Into the Twenty-First Century,* 25 Vand. J. Transnat'l L. (1992) at 225.

3.2.2 Legal Status of the Moon (and Other Celestial Bodies)

The most important concern at the time when the Moon Agreement was drafted was to ensure the peaceful and fair exploration and use of outer space.[17] The Moon Agreement is therefore designed to encourage resource development, with only major restriction imposed by the Treaty being the requirement that the Moon, other celestial bodies, and their resources be used for peaceful purposes only.[18]

One of the main provisions that define the legal status of the Moon and other celestial bodies is the non-appropriation principle. According to this principle, as expressed in both the Outer Space Treaty and the Moon Agreement,[19] the Moon is non-appropriable by means of sovereignty, occupation, or by any other means. This is the most fundamental relevant legal provision, which excludes the applicability of territorial sovereignty to outer space and any particular part thereof.[20] Outer space and its components do not form part of any State's territory, as legally defined for purposes of its sovereign authority; they do not constitute neither *res nullii* nor *terra nullii*, nor are they likely to be occupied in a legal manner.[21] The same is also applicable to the Moon, as it constitutes part of outer space. However, the legal status of mineral resources on the Moon and other celestial bodies is uncertain, particularly in terms of their commercial utilization. Such a consideration is very topical, as the status of extracted mineral resources and subsequent benefits is not explicitly defined by the space treaties.

3.2.3 Does Necessity Justify Failure?

In discussing about the failure or success of this treaty, it is essential to first examine the need for an agreement to govern activities on the Moon. In the author's view, this necessity stepped out of the relation between the Outer Space Treaty and lunar activities. When Apollo 11 landed on the Moon in 1969, the international community became fully aware that the general principles in the Outer Space Treaty were

[17] Moon Agreement, Art. 3, *supra* note 2, stating in para 1 that "[t]he moon shall be used by all States Parties exclusively for peaceful purposes" and in para 2 that "[a]ny threat or use of force or any other hostile act or threat of hostile act on the moon is prohibited. It is likewise prohibited to use the moon in order to commit any such act or to engage in any such threat in relation to the earth, the moon, spacecraft, the personnel of spacecraft or man- made space objects."

[18] Ibid.

[19] Article II, OST: "Outer space, including the Moon and other celestial bodies, is not subject to national appropriation by claim of sovereignty, by means of use or occupation, or by any other means."

Article 11(2), MA: "The Moon is not subject to national appropriation by any claim of sovereignty, by means of use or occupation, or by any other means."

[20] M. Lachs, The Law of Outer Space, (1972), at 42–48.

[21] For the concept of res nullius, see, e.g., I. Brownlie, *Principles of Public International Law* (Oxford: OUP, 1979), at 109, 180–1.

insufficient to regulate future exploitation activities on the Moon.[22] Additionally, the Outer Space Treaty did not offer detailed provisions on the involvement of private entities in space activities. As a result, the Moon Agreement was adopted in order to enhance and support the effectiveness of the Outer Space Treaty by setting forth principles and rules regulating the exploration, use, and exploitation of the Moon's and other celestial bodies' resources, until other treaties established more detailed relevant regulations.[23] The Moon Agreement does, indeed, constitute an elaboration on the Outer Space Treaty to the extent that it details the latter's fundamental rules with respect to the Moon and other celestial bodies in the solar system.[24]

However, the rather meagre acceptance of the Moon Agreement makes it difficult to argue any binding force of the rules provided by it, outside the limited number of aforementioned states party to it. It is, thus, evident that the binding legal value of the Moon Agreement cannot be compared to that of the Outer Space Treaty and that some provisions of the former may need to be amended to render the Agreement more appealing to States. Indeed, many concerns have been raised on the effectiveness and applicability of the Moon Agreement, as it is considered a "failed" treaty.[25] Therefore, the analysis of these rules as far as their legal status is concerned requires special care.

In the author's point of view, although the Moon Agreement enjoys minor support and has not been de facto effective since its adoption, it still constitutes the most appropriate legal text to start an analysis toward development of a regime governing commercial or private exploitation of the Moon and other celestial bodies. The Moon Agreement must be considered as an important development in international space law not only because it supplements some provisions and concepts of the Outer Space Treaty but also because it introduces new rules that could have practical impact on future missions on the Moon. For instance, Article 11 (3) of the Moon Agreement constitutes an elaboration of Article II of the Outer Space Treaty as it asserts that the Moon and other celestial bodies cannot be the property of private entities or even individuals, while at the same time, it allows for new rules to arise[26] and govern the possible establishment of manned stations on the Moon or environmental standards.

Comparing the Moon Agreement to the rest of the space treaties, the number of its signatures and ratifications imply the failure of a quite important legal text, which

---

[22] For example, this view was supported in van Bogaert ERC, Aspects of Space Law (Kluwer, Deventer), (1986), at 76.

[23] See United Nations Doc AIAC.1O5lS.R.187-S.R.188; J. A. Beesley, *Canadian practice in international law during 1972 as reflected mainly in public correspondence and statements of the Department of External Affairs,* 1 Canadian Yearbook of International Law, (1973), at 294–295.

[24] F.G. Von der Dunk, "Back in Business? The Moon Agreement, Private Actors and Possible Commercial Exploitation of the Moon and Its Natural Resources" (2006), paper presented at the International & Interdisciplinary Workshop on Policy and Law Relating to Outer Space Resources: The Example of the Moon, Mars & Other Celestial Bodies, Session 5 *The Acceptability of the Moon Agreement and the Road Ahead,* 28–30 June 2006, Montréal, Canada.

[25] Michael Listner, *supra* note 21.

[26] See Articles 4, 7, and 9, Moon Agreement, *supra* note 2.

should have been revised at least 20 years ago.[27] As numerous new realities of international and national space activities grow, such as commercialization of space and the appearance of new actors, the intensive need to pinpoint and revise, or amend, all those faulty provisions seems topical.

Coming back to the reasons of the marginal acceptance of the Agreement, it may be argued that such failure is due to certain problems concerning the scope of the Agreement's application, the lack of definition of celestial bodies, the concept of "common heritage of mankind" in article 11 (1, 5, 7), and last but not least, state responsibility arising from article 14, in relation to the Outer Space Treaty.

3.3 Legal Issues Rendering the Moon Agreement Ineffective

3.3.1 Scope of Application

The scope of application of the Moon Agreement is, according to the author, the first "thorn" in its applicability, resulting from the prolonged disagreement. Due to these controversies, States have been at first place reluctant and cautious in ratifying the Agreement.

Article 1 of the Moon Agreement provides for its applicability to the Moon and other celestial bodies, as long as no other more specific legal text exists to regulate such activities. According to its provisions, the Agreement "shall also apply to other celestial bodies within the solar system, other than the Earth, except insofar as specific legal norms enter into force with respect to any of these celestial bodies."[28] The scope of application of the Moon Agreement as defined in its first article was a cause of disagreement and discussion within COPUOS. The vital question was if its scope should extend to other galaxies and to circumlunar space.[29] As far as galaxies were concerned, and despite the insistence of the United States, COPUOS agreed that the Moon Agreement would be limited in scope in relation to the solar system.[30] On the other hand, States were reluctant as to circumlunar space, not so much with the concept per se, as mostly due to the lack of legal definition on what is "circumlunar space."[31] Eventually, COPUOS reached an Agreement on Article 1 referring to

[27] As set out by Article 18 of the Moon Agreement (MA), 10 years after its entry into force, COPUOS considered the question of a first review of the MA and the prospect of the establishment of an international regime at its 37th session in 1994. However, the Committee recommended to the General Assembly not to take further action at that time (GA Res. 49/34). This is the reason why a first review of the MA never took place.

[28] Article 1, Moon Agreement, *supra* note 2.

[29] Michael E. Davis, Ricky J. Lee, *Twenty Years After the Moon Agreement and its Legal Controversies, supra* fn. 4, at 11.

[30] See United Nations Doc A/AC.105/196.

[31] Nicholas M. Matte, *Legal Principles Relating to the Moon,* in Nandasiri Jasentuliyana and Roy S. Lee, Eds, *MANUAL ON SPACE LAW* (Springer: 1979), at 258.

orbits around or other trajectories to or around the Moon.[32] After a series of discussions on the interpretation of the treaty, COPUOS unanimously approved the Agreement on 14 December 1979.

3.3.2 *What Is a "Celestial Body"?*

The second problem identified during the early days of the debate in COPUOS, apart from the scope of application of the Agreement, was that of defining a "celestial body." According to one of the definitions adopted in space law literature, celestial bodies comprise the Moon, planets, moons, and asteroids (or planetoids) of our solar system, which are suitable for landing of manned or unmanned spacecraft, are of natural origin, and cannot be deviated from their celestial orbit.[33] In reality, and more specifically in astronomy, the concept of celestial bodies is much wider. However, from a legal point of view, the lawyer is interested only in celestial bodies that can become the scene of legally relevant actions.[34]

Due to the significant differentiation between bodies that may be objects of further exploration and exploitation, and others that are not appropriate for human space activities,[35] a more accurate legal definition should be adopted. Throughout the years, it has been argued that not one, but two legal regimes should be brought into existence, in order to regulate space objects. For some, there should be a distinction based on the existence of a human economic value, while for others, such a distinction would be contrary to the common interest principle already established in space law.[36]

In the author's opinion, there should be one basic foundation taken into consideration during the process and finalization of the definition of the term "celestial body": the legal status of the celestial bodies should not be distinct from that of outer space. Various international instruments have taken the same approach. The General Assembly of the United Nations in resolution 1962 (XVIII),[37] adopted in 1963, explicitly mentions the term "celestial bodies," allowing the right to explore and use them, and underlines the prohibition of their national appropriation. Therefore, no doubt remains that the legal status of celestial bodies is identical to

---

[32] Ibid.

[33] Gyula Gál, *SPACE LAW*, AW Sijthoff, Leyden (1969), at 186.

[34] *Ibid,* e.g., the sun, our solar systems, comets, etc. cannot become legally relevant for exploitation or exploration purposes.

[35] Inappropriate activities can be due to the object's dimension, nature, or substance: see G. P. Zhukov, *Kosmicheskoye arvo* (1966, Moscow; 1968, German ed., Weltraumrecht, Berlin), at 270–275; Marco G. Marcoff, *La lune et le droit international,* 68 Revue Generale de Droit International Public (1964), at 248.

[36] A. Gorbiel, *Remarques sur la definition de l' éspace extra-atmosphérique*, Proceedings of the 21st Colloquium on the Law of Outer Space (1978), at 89.

[37] General Assembly Resolution 1962 (XVIII), A/RES/18/1962.

the general status of outer space as defined by rules of international law and the "non-appropriation" principle. This should be taken as a starting point in order to establish a special rule, or rules, for the celestial bodies in conformity with the Outer Space Treaty and the Moon Agreement.

3.3.3 The Common Heritage of Mankind Principle

The Moon Agreement introduces in its 11th Article[38] the concept of the "common heritage of mankind," which refers to the Moon and its natural resources and calls for the establishment of an international regulatory regime "to govern the exploitation of the natural resources of the Moon."[39] The application of the common heritage of mankind concept to the exploitation of the natural resources of the Moon and other celestial bodies was one of the most debated and controversial topics during the negotiations of the Agreement.[40] Fundamentally, the Moon Agreement attributes to the Moon and other celestial bodies a form of common ownership that requires the compulsory sharing of advantages arising from any exploitation of celestial resources.[41] Such provision has been dysfunctional in relation to the general acceptance of the Moon Agreement, along with the rest of obligations imposed to States.[42]

Apart from the common heritage of mankind concept, Article 11 (3) also reiterates – and enlarges – the scope of the non-appropriation principle, with an explicit reference to private operators. The principle that outer space cannot be subject to the sovereign ownership and control of any State is not new and is presented in relation to Article II of the Outer Space Treaty,[43] which prohibits national appropriation by any means. Automatically, this bans any form of public ownership and sovereign rights, because celestial bodies are *res communis*.[44] However, the same does not apply to exploited materials, as *dominium* and ownership by a State could still be possible for private entities, creating thus intellectual property rights.[45]

The common heritage of mankind represents an evolution of the *res communis omnium* theory, by introducing an area not appropriable by States, which also has to

[38] Article 11 (1, 5, 7), Moon Agreement, *supra* fn. 2.

[39] *Ibid* (5).

[40] The Soviet Union opposed until the end of 1978, while the United States acted at first as one of the major supporters of the "Common Heritage of Mankind" principle during the negotiations. When the Agreement was open for signature and ratification, it strongly opposed it.

[41] Michael E. Davis, Ricky J. Lee, *supra* note 4, at 20.

[42] Ibid.

[43] Treaty on Principles Governing the Activities of States in the Exploration and Use of Outer Space, Including the Moon and Other Celestial Bodies, entered into force Oct.10, 1967, 18 U.S.T.2410, 610 U.N.T.S.205, Article II: *Outer space, including the Moon and other celestial bodies, is not subject to national appropriation by claim of sovereignty, by means of use or occupation, or by any other means.*

[44] Ian Brownlie, *Principles of Public International Law* (5th edition, Oxford University Press) (1998), at 105.

[45] Michael E. Davis, Ricky J. Lee, *supra* fn. 4, at 12.

be managed by all States acting collectively and exploiting it for the benefit of all.[46] Consequently, it transforms the *res communis omnium* theory into something that creates specific obligations on States that would have no sovereign role, yet act as representatives of all mankind. For instance, if a research financed by one State or a group of States took place, its results would be freely open and available to the international community[47]; however, this would be effective only in the event that these efforts contribute to the benefit of all mankind, otherwise they would be regarded as unsuitable.[48] This obligation has never been specifically defined as it relates to the extent of benefit sharing, contributing thus to States' unwillingness to sign and ratify the Moon Agreement. The major issue is that Article 11 has not indicated what the practical consequences of such obligations would be nor clarifies the boundaries between the exploration and use of the Moon on one hand and the Moon itself along with its natural resources on the other. There is a crucial distinction between the legal status of the Moon itself (or other celestial bodies), which is non-appropriable by States, and the legal status of natural resources, which is uncertain and particularly fruitful for commercial exploitation. Simply put, according to doctrine, the solution would be the adoption of an international legal regime providing full legal ownership to States that would share any profits derived from mineral exploitation in outer space, with special consideration to the interests of developing countries.[49]

Such a regime resembles the 1982 UN Convention of the Law of the Sea (UNCLOS),[50] which declared the deep seabed to be the common heritage of mankind, allowing States to regulate its mineral exploration and exploitation through the establishment of an International Seabed Authority.[51] According to the aforementioned convention, a private enterprise must be licensed by the International Seabed Authority and return to the latter a certain percentage of its profit, as well as to transfer the technology used on a *fair and reasonable commercial basis*,[52] provision

[46] M. Lucy Stojak, *Rapporteur's Notes for Session 5 The Acceptability of the Moon Agreement and the Road Ahead* (Commentary on F.G. Von der Dunk's, *Back in Business? The Moon Agreement, Private Actors and Possible Commercial Exploitation of the Moon and Its Natural Resources* (2006), *supra* note 30), at 314.

[47] Article 3, Moon Agreement, *supra* note 2.

[48] Michael E. Davis, Ricky J. Lee, *supra* note 4, at 20.

[49] B. A. Boczek, *Ideology and the law of the sea: the challenge of the new international economic order*, 7 Boston College International and Comparative Law Review 1 (1984); see C. Q. Christol, *The common heritage of mankind provision in the 1979 Agreement Governing the Activities of states on the Moon and other Celestial Bodies*, 14 International Lawyer (1980), at 429; M. A. Ferrer, *Legal implications of the principle according to which exploration and use of outer space shall be carried out for the benefit and interest of all states, taking into particular account the needs of developing states*, Proceedings of the 32nd Colloquium on the Law of Outer Space 5 (1989).

[50] United Nations Convention on the Law of the Sea, 1833 UNTS 3; 21 ILM 1261 (1982) [UNCLOS].

[51] Articles 151–161, 1982 Law of the Sea Convention, *supra* fn. 57.

[52] *Ibid,* Article 170.

opposed by developed States, such as Germany, France, the United States, and the United Kingdom.[53] In an analogy, the same would happen, had such a principle been introduced to the Moon Agreement or otherwise been imposed to the legal regime governing Moon exploitation. States of the developed world would never accept to invest in mining ventures or manned missions, be bound by obligations unfavorable to their interest, and allow developing States take advantage of their efforts. A similar scenario took place in the case of seabed mineral exploitation eventually leading to the revision of the UNCLOS in 1994.[54]

Therefore, the application of the common heritage of mankind concept to the natural resources of the Moon within the context of the Moon Agreement would require a different approach, or even total withdrawal from the text of the Agreement itself, as it is a concept applicable to only 17 States, none of which has immediate plans or funds to host a lunar exploitation mission. Besides this, the fact that the concept applies only to the resources of the celestial bodies prior to their removal, meaning thus that full ownership may be exercised over them after extraction takes place notwithstanding the provisions of the Moon Agreement,[55] does not seem cohesive.

3.3.4 Responsibility of States for Space Activities

Article 14 of the Moon Agreement, which builds and further elaborates on Article VI of the Outer Space Treaty, provides for States to bear international responsibility for national activities on the Moon, even if these activities are undertaken by nongovernmental entities.[56] Such a concept, as introduced in the Outer Space Treaty,[57] seems to create specific obligations imposed on States,[58] which are, however, not further explained in the Moon Agreement, raising thus important issues. According to the aforementioned responsibility regime, States shall bear interna-

[53] Agreement Concerning Interim Arrangements Relating to Polymetallic Nodules of the Deep Sea Bed, Sept. 2, 1982, 34 U.S.T.S. 3451.

[54] "[t]o address certain difficulties with the seabed mining provisions contained in Part XI of the Convention, which had been raised, primarily by the industrialized countries, the Secretary-General convened in July 1990 a series of informal consultations which culminated in the adoption, on 28 July 1994, of the Agreement relating to the implementation of Part XI of the United Nations Convention on the Law of the Sea of 10 December 1982. The Agreement entered into force on 28 July 1996," *available at* http://www.un.org/depts/los/convention_agreements/convention_overview_part_xi.htm.

[55] United States Senate Committee on Commerce, Science and Transportation, 96th Congress, 2nd Session, Report on Agreement Governing the Activities of States on the Moon and Other Celestial Bodies (1980), at 30.

[56] Article 14, Moon Agreement, *supra* fn. 2.

[57] Article VI, Outer Space Treaty, *supra* fn. 14.

[58] Bin Cheng, *Article VI of the 1967 Space Treaty revisited: 'international responsibility', 'national activities' and 'the appropriate nation'*, 26 Journal of Space Law 7 (1998), at 12–19.

tional responsibility for "national activities" on the Moon and shall be held liable in conformity with the relevant provisions of the international space treaties referred to. These provisions ensure the accountability of States at the international level for private and national commercial activities on the Moon,[59] the latter being part of outer space.

To begin with, the notion of "state responsibility" in itself is quite unclear in the context of space activities. States are required to ensure compliance of their national space activities with the obligations which arise under the relevant space treaties and to also provide authorization and supervision to nongovernmental entities.[60] At a first glance, direct space responsibility is assumed, meaning that activities undertaken by a nongovernmental entity are still attributable to the State, as if it were an act of its own. Consequently, a State would be responsible for any breach of international law and as such responsible for not "assuring" compliance by nongovernmental entities.[61] It is all a matter of interpretation of Article VI of the Outer Space Treaty along with Article 14 of the Moon Agreement; States are not responsible to one another for breaches of their national laws caused by their nongovernmental entities. Such a vague interpretation emerges, though, from the fact that States would be indirectly responsible for acts of private entities under their effective jurisdiction.[62]

The notion of "responsibility" in the aforementioned context is also closely linked to the term "national activities" which equally lacks adequate definition. A national space activity could either be considered as an activity realized by the State of Registry of a space object, or alternatively, a "national" space activity could be defined based on the nationality of the persons involved.[63] However, both these approaches seem quite ineffective in the context of Article 14 of the Moon Agreement. Firstly, if international responsibility for national activities depended solely on registration, only the State of Registry would be responsible,[64] allowing thus other participating States that have not registered the object to evade responsibility. Secondly, if responsibility for national space activities was based on the nationality of persons involved, it would be problematic for nationals operating from foreign territories, causing States to assume responsibility beyond their effective jurisdiction.[65] However, it is the author's contention that it would be preferable to accept the interpretation of the term "jurisdiction" in Article 14 of the Moon Agreement as one indicating that States should bear international responsibility for

---

[59] F.G. Von der Dunk, *Back in Business? The Moon Agreement, Private Actors and Possible Commercial Exploitation of the Moon and Its Natural Resources* (2006), *supra* fn. 30, at 246.

[60] Article VI, OST, *supra* fn. 14.

[61] Bin Cheng, *Article VI of the 1967 Space Treaty revisited: 'international responsibility', 'national activities' and 'the appropriate nation'*, *supra* fn. 65, at 16.

[62] *Ibid,* at 18.

[63] Article IX, Outer Space Treaty, *supra* fn. 14.

[64] Convention on Registration of Objects Launched into Outer Space, *entered into force* Sept. 15, 1976, 28 U.S.T. 695, 1023 U.N.T.S. 15., Article II (2).

[65] Bin Cheng, *The extra-terrestrial application of international law,* 18 Current Legal Problems (1965), at 132.

any kind of activities within their legal sphere of authority. As a result, any activity, be it carried out by governmental or nongovernmental entities within a nation's territorial, quasi-territorial, and personal jurisdiction, would qualify as "national activity."

Unfortunately, it is the author's view that the Moon Agreement does not adequately confirm the validity of principles concerning State responsibility as found in the Outer Space Treaty and international law. On the contrary, it seems that it creates new obligations for States that are not defined by ascribing State responsibility within States' territory. This is one more legal gap of the treaty confirming its lack of effectiveness, as well as its controversies and the lack of acceptance by States.

3.4 The Future of the Moon Agreement: Make It or Break It?

The past and the present have shown that the Moon Agreement has not been as effective in practice as it was initially hoped. Its lack of cohesion in relation to the rest of the space treaties and some major legal impediments, as earlier analyzed, have turned the Moon Agreement into the "outcast" of the space legal regime.

The question that follows would be how to guarantee the future of the Moon Agreement given that in recent years a renewed interest in exploring and exploiting celestial bodies has been observed together with the fact that a number of developed States fund resource exploration and exploitation missions. The problem, however, is that these developed States and mostly the "Big Three" (the United States, Russia, and China) have not ratified the Agreement. The fact that the legal framework of the Outer Space Treaty is the only one that covers these missions automatically bans the supplementary function of the provisions of the Moon Agreement at the level of national space activities. In the last 5 years, the LSC and UNCOPUOS have paid particular attention to the low ratification of the Agreement. This led to the 2008 Joint Statement on the benefits of adherence to the Agreement Governing the Activities of States on the Moon and Other Celestial Bodies of 1979,[66] in which States parties to the Agreement pointed out that "the Agreement does not preexclude any modality of exploitation, by public and/or private entities, nor forbids commercial treatment, as long as such exploitation is compatible with the requirements of the Common Heritage of Mankind regime." Additionally, it ensures States that are not yet parties to the Agreement that this legal text constitutes a "mutual commitment to seeking a multilateral solution." However, the statement was ineffective as the situation continued to be a "dead end" with none of the spacefaring nations willing to ratify the Agreement.

As a result, a "catchup" of the Agreement with the rest of the space treaties, especially with the Outer Space Treaty, is of vital importance. Article 18 of the

[66] See Doc A/AC.105/C.2/2008/CRP.11.

Agreement provides that 10 years after its entry into force, states party have the possibility to revise it. Such revision has not been materialized until today. In the author's opinion, should all signatory States withdraw from the Moon Agreement, crucial legal gaps will arise. This will be especially true if the Outer Space Treaty is not updated to include in its scope activities currently covered by the Moon Agreement.

An amendment of the Moon Agreement could be the first step to renew its spirit and purpose, as it seems to be the only feasible existing starting point for the establishment of a fair and efficient international legal regime for the exploitation of celestial bodies' resources. However, this will not be an easy task as the amendment procedure could prove to be a cumbersome one as it will have to accommodate new interests and needs of modern space actors.

In more details, the first major impediment that should be considered is no other than the common heritage of mankind concept of Article 11. The review of the common heritage of mankind concept should be done in regard to the commercial exploitation and use of the Moon and other celestial bodies by providing details as to how such exploitation could be feasible, especially with regard to space mining. If this principle remains as is, it will continue to oppose the "province of all mankind" concept[67] and impede discussions on future rules, parameters, and procedures that should be established to strike a balance between public and private interests. However, the author doubts whether the common heritage of mankind concept should be removed or replaced by the "province of mankind" concept or even attributed a more specific meaning within the text of the Agreement. If the common heritage of mankind concept is maintained in case of revision, it should be clarified – based, for instance, on the similar provisions that exist in the context of the law of the sea.

One more possible scenario would be the implementation of the mechanisms provided for in Article 11(5) of the Agreement with the establishment of an international regime. The immediate establishment of an international regulatory regime will set out the principles and terms of commercial exploitation of mineral resources and the grounds for the powers and functions of a regulatory body. For example, a regime similar to that provided in Part XI of the UNCLOS (as revised in 1994) could be introduced to address the problem of Moon Agreement's ineffectiveness. In order to enhance the commercial aspect of lunar missions, there could be an

[67] Article 4, Moon Agreement, *supra* fn. 2 stating in para 1 that "[T]he exploration and use of the Moon shall be the province of all mankind and shall be carried out for the benefit and in the interests of all countries, irrespective of their degree of economic or scientific development. Due regard shall be paid to the interests of present and future generations as well as to the need to promote higher standards of living and conditions of economic and social progress and development in accordance with the Charter of the United Nations" and in para 2 that "[S]tates Parties shall be guided by the principle of cooperation and mutual assistance in all their activities concerning the exploration and use of the Moon. International cooperation in pursuance of this Agreement should be as wide as possible and may take place on a multilateral basis, on a bilateral basis or through international intergovernmental organizations."

international authority (such as the International Seabed Authority "ISA" in law of the sea) empowered to control access to the celestial bodies and grant approval to States. The authority would also be able to distribute the benefits and decide on the transfer of technology from developed to developing States.[68] Yet, it is crucial to determine whether the requirement that the exploration and use of outer space be "for the benefit and in the interest of all countries" imposes a positive and specific obligation "regarding the sharing the benefits of space exploration and use" or merely constitutes an "expression of desire that the activities should be beneficial," in contrast to being harmful "in a general sense."[69]

3.5 Concluding Remarks

Regardless of the path that lawmakers will follow, the scope of the Moon Agreement is more relevant than ever. The Agreement should thus remain as a legal text but be either revised or amended to answer present and future needs. As already noted, this will not be an easy task; the major challenge is the acceptance by all States that are party to the Outer Space Treaty. Imperfect as it may be, the Moon Agreement still represents the most adequate starting point for developing a regime to strike the balance between public and private interests with respect to any commercial activities on the Moon and other celestial bodies. Unfortunately, reinterpretations by means of non-binding documents, statements, or suggestions have never proved to suffice. As the Moon and other celestial bodies constitute a promising part of business, an updated version of the Moon Agreement within the wider scope of the unification of the space treaties should be considered as soon as possible to enable humanity's presence in and exploration of such bodies.

---

[68] Alan Duane Webber, Note, *Extraterritorial Law on the Final Frontier: A Regime to Govern the Development of Celestial Body Resources*, 71 Geo. L.J. 1427, 1436–37 (1983); See, e.g., International Space Activities, 1979: *Hearings Before the Subcomm. on Space Science & Applications of the H. Comm. on Science & Technology*, 96th Cong., (1979), at 134.

[69] Gorove, *Implications of International Space Law for Private Enterprise,* 7 Annals Air & Space Law 319, (1982), at 321.

Chapter 4
The Interplay Between Space Law and International Investment Law: Local Equity Requirements as a Model for Standards of Global Action in the Uses of Outer Space

Lukas Vanhonnaeker and Maria Manoli

4.1 Introduction

Private investment in the realm of space activities is an ever-growing reality. Only in the first quarter of 2018, more than US $1 million was invested by the private sector in start-ups in the space industry.[1] As a result, space technologies are developing fast,[2] national laws are trying to adjust to the new realities,[3] and traditional space actors appear in different roles.

The involvement of the private sector in the field of space activities is not new. It has existed since the first space activities and contributed to the advancement of space technology and, subsequently, to the improvement of human life on Earth.[4] So far, the private space sector has harmoniously cooperated with the public sector at both the

[1] Space Angels, "Space Investment Quarterly: Q1 2018 Report" (10 April 2018), online: Space Angels https://www.spaceangels.com/post/space-investment-quarterly-q1-2018.

[2] See, for example, "Incredible Technology – Space Travel and Exploration," online: https://www.space.com/topics/incredible-space-tech.

[3] See *US Commercial Space Launch Competitiveness Act*, Pub L 114–90 (2015), and *Luxembourg Draft Law on the Exploration and Use of Celestial Bodies*, Le Gouvernement du Grand-Duché de Luxembourg, Ministère de l'Économie (2015), online, Government of Luxembourg: http://www.gouvernement.lu/6481974/Draft-law-space_press.pdf.

[4] See generally, John Adolph, "The Recent Boom in Private Space Development and the Necessity of an International Framework Embracing Private Property Rights to Encourage Investment" (2006) 40:4 *The International Lawyer* 961–985.

L. Vanhonnaeker
Faculty of Law, McGill University, Montreal, QC, Canada

M. Manoli (✉)
Institute of Air and Space Law, Faculty of Law, McGill University, Montreal, QC, Canada
e-mail: maria.manoli@mail.mcgill.ca

G. D. Kyriakopoulos, M. Manoli (eds.), *The Space Treaties at Crossroads*,
https://doi.org/10.1007/978-3-030-01479-7_4

national and international levels.[5] Over the past decade, however, the private space sector has demonstrated an unprecedented potential for a new era of space activities, with the main one being plans for the exploitation of mineral resources from the celestial bodies of outer space.[6] The projected benefits from such activities constitute a particularly appealing incentive for states[7] that adapt their laws to satisfy the objectives of the private space sector and create a safe environment for private space companies.[8]

However, the governance and regulation of the uses of outer space is not a national matter. As part of our global commons, outer space is characterized by a globality that requires the involvement of all states and other stakeholders in both the regulation of the uses of outer space and its use per se.[9] In this context, and given that the participation of the private space sector is essential for the development of space activities, the question arises as to how harmoniously coordinate the development of space activities through the involvement of the private space sector without disregarding the concept of common interest that is inherent to the five UN Space Treaties.[10] This paper attempts to use the case of international investment law, whose

---

[5] Ibid.

[6] See, for example, the proposed space mining plans of private companies such as: Moon Express, more information herein, online, Moon Express http://www.moonexpress.com/expeditions/; Shackleton Energy, more information herein, online, Shackleton Energy http://www.shackletonenergy.com/; Planetary Resources, more information herein, online, Planetary Resources http://www.planetaryresources.com/asteroids/#asteroids-intro; Deep Space Industries, more information herein, online, http://deepspaceindustries.com/mining/.

[7] Jeff Desjardins, "The Value of Asteroid Mining" (3 November 2016), online: Business Insider https://www.businessinsider.com/the-value-of-asteroid-mining-2016-11.

[8] See *US Commercial Space Launch Competitiveness Act* and *Luxembourg Draft Law on the Exploration and Use of Celestial Bodies*, *supra* note 3.

[9] See, for instance, B Cheng, *Studies in International Space Law* (Oxford: Clarendon Press, 1997); F Lyall, P B Larsen, Space Law – A Treatise (UK: Ashgate, 2009); G Gal, *Space law* (Leiden: Sijthoff, 1969); I H P Diederiks-Verschoor, Vladimír Kopal, *An introduction to space law* (Alphen a/d Rijn: Kluwer Law International, 2008); I A Vlasic, Ed, *Explorations in aerospace law: selected essays by John Cobb Cooper*, (Montreal: McGill University, 1968); Manfred Lachs, *The Law of Outer Space: An Experience in Contemporary Law-Making* (The Netherlands: Martinus Nijhoff Publishers, 2010); Marietta Benko, Kai-Uwe Schrogl, Eds, *Space Law: Current Problems and Perspectives for Future Regulation*, Essential Air and Space Law (The Netherlands: Eleven, 2005); Nandasiri Jasentuliyana, Roy S K Lee, Eds., *Manual on space law. Volume I*, (Dobbs Ferry: Oceana, 1979).

[10] *Treaty on Principles Governing the Activities of States in the Exploration and Use of Outer Space, including the Moon and other Celestial Bodies*, of 27 January 1967 (610 UNTS 206) ["Outer Space Treaty"]; *Agreement on the Rescue of Astronauts, the Return of Astronauts and the Return of Objects Launched Into Outer Space*, 22 April 1968, 672 UNTS 119, 19 UST 7570, TIAS No 6599, 7 ILM 151 (entered into force 3 December 1968) ["Rescue Agreement"]; *Convention on International Liability for Damage Caused by Space Objects*, 29 March 1972, 961 UNTS 187, 24 UST 2389, 10 ILM 965 (1971) (entered into force 1 September 1972) ["Liability Convention"]; *Convention on Registration of Objects Launched into Outer Space*, 6 June 1975, 28 UST 695, 1023 UNTS 15 (entered into force 15 September 1976) ["Registration Convention"]; and *Agreement governing the Activities of States on the Moon and Other Celestial Bodies*, 5 December 1979, 1363 UNTS 3 (entered into force 11 July 1984) ["Moon Agreement"].

concepts can be adopted and adjusted to address the specificities of the emerging private space industry.

Like any kind of development, development through foreign investment did not come free from sacrifices. It required mutual concessions and harmonious cooperation between the host state (state where the investment takes place) and the home state (state of the investor), both being necessary elements for a conflict-free environment.[11] Bilateral agreements seek to achieve this purpose. As a result of negotiations, bilateral investment agreements allow for a careful crafting of provisions aiming to promote mutual interests as well as to impose limits.[12] The interconnectivity and interdependence that define our globalized world were the main reasons behind this necessity and the main realities that allowed foreign investments to flourish.[13]

Similar characteristics can be observed in the domain of investments with the use of outer space as their object. However, in this case, the interconnectivity and interdependence do not take place at the transnational or international levels. Rather, they are the result of the global nature of outer space as a resource, nature that derives from the legal and natural status of outer space as free from private and public exclusivity.[14] Therefore, as part of our global commons, outer space resources call for a global rather than bilateral or multilateral or even international understanding. It remains to see how a regime mostly based on bilateralism can create an example for one that requires a global approach.

4.2 The Different and Yet Similar Origins of the Two Regimes

International investment law is composed of a *corpus* of bilateral and multilateral agreements that aim to protect and promote foreign direct investment ("FDI"). It is composed of a web of more than 3300 agreements[15] initially entered into mostly between developing countries and developed counties. In fact, international invest-

---

[11] See, for instance, Junji Nakagawa, *Nationalization, Natural Resources and International Investment Law – Contractual Relationships as a Dynamic Bargaining Process* (London/New York: Routledge, 2018), at 15.

[12] See, in general, Rudolf Dolzer and Margerete Stevens, *Bilateral Investment Treaties* (The Hague: Martinus Nijhoff/ ICSID, 1995), Ch. 1.

[13] Debra P Steger, "International Trade and Investment: Towards a Common Regime?" in Roberto Echandi and Pierre Sauvé, Eds, *Prospects in International Investment Law and Policy: World Trade Forum* (Cambridge: Cambridge University Press, 2013), at 156–165.

[14] For the concept of exclusivity, see, generally, Philip de Man, *Exclusive Use in an Inclusive Environment: The Meaning of the Non-Appropriation Principle for Space Resource Exploitation*, Space Regulations Library (Switzerland: Springer, 2016).

[15] UNCTAD, *World Investment Report 2018: Investment and New Industrial Policies* (UNCTAD, 2018), at 88 [footnote omitted]:

ment law was born in a tensed historical context characterized by a strong dichotomy between the North (developed countries) and the South (developing countries). Indeed, before the emergence of international investment law as we know it, investors, often originating from developed countries, would remain under the jurisdiction of their home state, and their consuls would exercise jurisdiction over them in the host state, often a developing country. Such a rule remained applicable and easily enforced by colonial powers in developing countries.

However, with the advent of the nation-state system and the general movement of decolonization, colonial territories began to gain their independence and started to challenge the practice according to which foreign investors remained subject to their own jurisdiction even in foreign territory, thus undermining the doctrines of sovereignty and sovereign equality. As a result, the notions of "national treatment" and "minimum standards of treatment" were born together with principle according to which foreign investors "were not entitled to any greater protection than those accorded to the nationals of the country under the law of the land."[16] However – what was soon to become the crux of the issue – the "international minimum standard" was not defined. Developing and developed countries could not reach a consensus, the former being firmly attached to the principle according to which investors and investments should be subject to the domestic law of the host state and the latter strongly advocating the application of a minimum standard of treatment consisting of protective standards of protection based on customary international law emanating from the practice of developed countries.

In an effort to provide a legal framework characterized by legal certainty, series of attempts were made to reach a multilateral agreement[17] regulating FDI through negotiations gravitating around the "classic doctrinal tension that existed [and arguably still exists] with regard to foreign investment law (…) between those who preferred to apply national law to regulate foreign investment and those who preferred to apply international law, meaning customary international law made by investor countries."[18] These attempts culminated with the negotiation of the Multilateral Agreement on Investment (MAI) that ultimately failed.

After the failed attempts to reach a common understanding on how to regulate FDI, developing countries had to face important economic difficulties in the 1980s, and the tide began to turn, leading them to realize that FDI would be part of the solution to enhance their economies. They thus started not only to reverse their traditional skeptic attitude toward FDI, but they even engaged in a "race" to attract foreign

In 2017, countries concluded 18 new IIAs: 9 bilateral investment treaties (BITs) and 9 treaties with investment provisions (TIPs). This brought the size of the IIA universe to 3322 agreements (2946 BITs and 376 TIPs), of which 2638 were in force at year-end.

[16] Surya P Subedi, *International Investment Law* (Sussex: Hart Publishing, 2008), at 8. In addition, it was submitted that if the local law was considered inferior, not well developed or failed to meet the standards of justice and equity, the international minimum standard rather than national law would apply to foreign investors (*Ibid*, at 9).

[17] *Ibid*, at 11 *et seq*.

[18] *Ibid*, at 10.

investors, reaching the point where they would offer them unprecedented incentives. The process of entering into bilateral investment treaties (hereafter "BITs") and free-trade agreements (hereafter "FTAs") with investment chapters had begun. By concluding such agreements, developed countries managed to ensure the protection of their investors and developing countries that had in the past gone down the road of expropriation and nationalization of foreign investments now accepted such provisions through BITs and other regional treaties to create a stable investment climate. In other words, developing countries, in the perspective of enhancing economic development by attracting foreign investments, decided to limit their discretionary power with respect to the treatment of foreign investors investing in their territory.

While the development of modern international investment law is closely linked to the dynamic and context of the colonial era and to the decolonization process, the origins of international space law find their roots in a different setting, yet sharing common characteristics. Built upon a spirit of anti-imperialism, space law set the basis for an anticolonial exploration of outer space.[19] From the outset, space law's creation was based on the fear of interstate domination and subjugation as it was developed during the environment of the Cold War era.[20] Taking as lessons the devastating consequences of WWII to the autonomy of territories and to the peace of the world, space law, from its inception, had no other objective than to promote global advancement through the uses of outer space as well as achieve peace and international cooperation.[21]

In other words, while international investment law departed from unilateralism in order to achieve the necessary extent of cooperation, mutual understanding, and mutual benefit – even if initially mostly at a bilateral level – space law was ab initio constructed on the idea of globality.

Therefore, investment law's bilateralism can be understood as the merging of two localities. Originating from a home state (the state of the investor – *locality A*), the investment operation is destined to take place in the host state (the state where the investment takes place – *locality B*). International investment law's biggest challenge was thus to accommodate, mitigate, and harmonize the interests of two actors or of two localities in addition to those of the foreign investors. Hence, bilateralism offered a significant opportunity to the parties involved: the privilege of defining the behavior of locality A with respect to the investors of locality B and vice versa. As such, international investment law did not manage so far to achieve multilateralism

[19] Virgiliu Pop, *Who Owns the Moon?: Extraterrestrial Aspects of Land and Mineral Resources Ownership* (New York: Springer, 2009), at 61.

[20] Boutros Boutros-Ghali, "International Cooperation in Space Activities for Enhanced Security in the Post-Cold War Era" in Karl-Henz Bockstiegel, et al., Eds, *Space Law – Basic Legal Documents – Volume I* (The Netherlnads: Eleven International Publishing, 1991), at 1.

[21] *Outer Space Treaty*, *supra* note 10, art III:

> States Parties to the Treaty shall carry on activities in the exploration and use of outer space, including the Moon and other celestial bodies, in accordance with international law, including the Charter of the United Nations, in the interest of maintaining international peace and security and promoting international cooperation and understanding.

at the global level, partly because it is a field of law that heavily relies on the notion of borders, the concept of different territories, and of different – and often conflicting – interests.

On the other hand, the global nature of outer space required, from the beginning, a stricter approach to what such *localities* would be and what they would mean for the world. As a result, commonly agreed principles and rules were adopted through the means of multilateralism, which ensured to promote, among other, the "common interests of all mankind,"[22] the "benefit of all peoples,"[23] the "benefit and interests of all countries,"[24] the "international peace and security,"[25] and the "international cooperation and understanding."[26] Accordingly, unlike the ad hoc and case-specific provisions of investment treaties, space law, and in particular the Treaty on Principles Governing the Activities of States in the Exploration and Use of Outer Space, including the Moon and Other Celestial Bodies ("Outer Space Treaty"), regulated the common investment of all humankind in outer space for its own betterment. Therefore, in the realm of international investment law, the two different localities, respectively, that of the investor (*locality A*) and that of the investment (*locality B*), crystallize their collaboration in investment treaties for the good of a particular transnational activity. On the contrary, in the realm of space law, the "locality" of humankind and the "locality" of investment are identical (outer space), thus excluding bi- or unilateralism.

4.3 Humankind as the Home and Host State of Investments

International investment agreements are key for investor as they signal a stable and predictable investment environment. More than providing "foreign investors with powerful new rights to protect their investments against expropriation and other forms of discrimination and the ability to sue governments directly through an innovative form of dispute settlement known as investment treaty arbitration,"[27] international investment agreement insures some degree of legal certainty.

Indeed, in order for investors to engage in economic operations in foreign territories, they need incentives. Investors are typically private entities or government actors acting as private investors (i.e., state-owned entities or sovereign wealth funds (SWF)) that are profit-driven and usually not in the perspective of engaging in nonprofit activities. Accordingly, investors will typically engage in operations that involve the lowest degree of risk possible in terms of transaction costs. One of these

[22] *Outer Space Treaty*, *supra* note 10, Preamble.

[23] *Ibid.*

[24] *Ibid*, Art I.

[25] *Ibid*, Art III.

[26] *Ibid.*

[27] Eric Gottwald, "Leveling the Playing Field: Is it Time for a Legal Assistance Centre for Developing Nations in Investment Treaty Arbitration?" (2007) 22 *Am. U. Int'l L. Rev.* 237, at 238–239.

transaction costs consists of the uncertainty associated with the operation at issue which encompasses the certainty provided by the applicable legal regime, or, for that matter, the uncertainty of investing in a context characterized by a non-exhaustive or nonexistent legal regime; the absence of legal certainty would indeed act as a hindrance to investments.

Unlike international investment law, where the alleviation of uncertainty takes place through the negotiation of investment agreements, space law does not provide for such a mechanism. Rightly so, the governance and use of outer space is not a negotiable matter, and, consequently, it cannot be regulated by one, two, or few states. Its global nature requires global treatment. However, despite the differences between the two regimes, certain international investment law mechanisms and concepts can be useful to address the specificities of investments having for object the use of outer space.

The main difference between FDI and investments taking place in the context of a global commons, such as outer space, lies within the local character of FDI, which is absent from investments with respect to global commons. FDI "involves the transfer of tangible or intangible assets from one country to another for the purpose of their use in that country to generate wealth under the total or partial control of the owner of the assets."[28] Outer space investments could not qualify as such not only due to the lack of the element of transfer between two states but also because outer space does not have an owner. Therefore, the question arises as to how investment can take place when it involves the use of outer space by private entities and under what circumstances.

The answer to this question lies within the global nature of space law and its provisions. First and foremost, a careful reading of the declaratory provisions of the Outer Space Treaty[29] reveals that outer space cannot constitute a source of wealth for the owner of a specific investment but for all humankind, as it is the *exploration* and *use* of outer space that is meant to be "carried out for the benefit and in the interests of all countries" and not the material dimension of outer space per se.[30] Similarly, it is the same *exploration* and *use* that is considered to be the "province of all mankind," meaning that the use of outer space and its fruits must be savored by all humankind.[31]

The same logic is further evidenced through the responsibility regime for outer space activities, according to which states bear international responsibility for public and private space activities and are also responsible for authorizing them and continuously supervising them[32]:

---

[28] M. Sornarajah, *The International Law on Foreign Investment* (3rd ed, Cambridge: Cambridge University Press, 2010), at 8

[29] *Outer Space Treaty*, *supra* note 10, Preamble, arts I, II, and III

[30] *Ibid*, art I.

[31] See "Space Mining and the Use of Space Natural Resources" in Ram S Jahku and Joseph N Pelton, *Global Space Governance: An International Study* (Switzerland: Springer, 2017), at 379–413.

[32] *Outer Space Treaty*, *supra* note 10, art VI.

> States Parties to the Treaty shall bear international responsibility for national activities in outer space, including the Moon and other celestial bodies, whether such activities are carried on by governmental agencies or by non-governmental entities, and for assuring that national activities are carried out in conformity with the provisions set forth in the present Treaty. The activities of non-governmental entities in outer space, including the Moon and other celestial bodies, shall require authorization and continuing supervision by the appropriate State Party to the Treaty …[33]

Therefore, whereas in the case of international investment law, the state seeks to achieve favorable circumstances for the investments of its private investors in a foreign jurisdiction, in the case of space law, the state must ensure that the behavior of its private entities complies with the purposes of the global framework, i.e., the *exploration* and *use* of outer space in the interests of all countries.

Regardless of how different the role of states in international investment law and space law may be, the latter can be useful to craft a regulatory framework for the former. To do so, the investor undertaking space activities would have to act under the direction of its own state. The state would thus direct the private actor's behavior to ensure compliance of the activities with the interests of all states and those of all humanity.

Furthermore, FDI is profitable for the social and economic development of host states, for example, by creating new jobs and via transfers of technology and generally for the wealth of the economy. However, regulating FDI via international investment agreements can also have negative impacts, such as on states' sovereign regulatory power: abiding by their international obligations necessarily entails limiting the ability of states to regulate domestically.

Similarly, private space investments can be profitable for the social and economic development of humankind and humanity. However, more importantly, the negative implications that investment agreements can have on the sovereign regulatory power of a host state are eliminated in the case of outer space investments. This is because the interests of the home state (the state of investor) align with those of humankind and humanity, even if it is because of their international obligations.

4.4 Local Equity Requirements in International Investment Law as an Example for Minimum Standards of Global Action in the Uses of Outer Space

Local equity requirements, also known as "joint venture requirements," are a well-known concept in international trade and investment law, and they refer to the requirement for the foreign investor to acquire domestic equity in order to enter a host state market. Such requirements generally provide that the foreign investor cannot hold a majority stake in a domestic company in order to guarantee that the control of the economic operation rests with the host state or the local partner.

[33] *Ibid.*

In the case of outer space investments, "local" equity requirements would necessitate that the investment is undertaken by a company, whose majority stake is owned by a state.[34] In FDI the rationale behind local equity requirements is to ensure that the state preserves a control over the foreign investment. In the context of outer space investments, local equity requirements would fulfill the principles and objectives of space law. Moreover, the application of such standards would also preserve the nature of outer space as non-exclusive and free from appropriation and sovereignty as the use of outer space would be governed through the means of a collective rather than mono- or oligopolistic scheme.

In international investment law, local equity requirements or, for that matter, any entry requirement are not illegal and do not need to be justified by the host state. The sovereignty of states dictates that they have complete control over the entry and movement of capital within their territory[35]:

> The general principle of sovereignty over national territory implies that each State is able to determine which foreign investments will be allowed in and which it may refuse. Neither international trade law nor international investment law, which deal directly or indirectly with investment issues, require that a State allow the entry of a foreign investment into its territory against its will.[36]

Within the same logic, the absence of sovereignty in outer space and the status of its exploration and use as in the interests of all countries would *imply that no state alone would be able to determine which investments will be allowed in and which it may refuse*.

Examples of such requirements can be found in Article 7 of the Montenegro Foreign Investment Law[37] which provides

> Foreign investor may invest in a domestic business entity that manufactures and trades in armaments, ammunition or explosives, or establish such a business entity in the Republic with a domestic legal or natural person, as the case may be.
>
> The share of the foreign investor in the stock capital or property rights *or voting rights* in the business entity referred to in paragraph 1 of this Article shall not exceed 49%.

Article 6 of the Guinean Investment Code[38] provides that

> Les personnes physiques ou morales de nationalité étrangère ne peuvent détenir, directement ou à travers des sociétés de droit guinéen, plus de 40% des titres sociaux d'entreprises engagées en Guinée dans les activités suivantes :
> - la publication de quotidiens ou de périodiques d'information générale ou politique ;
> - la diffusion de programmes télévisés ou radiophoniques.

---

[34] Talkmore Chidede, *Legal Protection of Foreign Direct Investment – A Critical Assessment with Focus on South Africa and Zimbambwe* (Hambourg: Anchor Academic Publishing, 2016), at 72.

[35] See Jeswald W Salacuse, *The Three Laws of International Investment: National, Contractual, and International Frameworks for Foreign Capital* (Oxford: Oxford University Press, 2013), at 75–88.

[36] Armand de Mestral, "Pre-Entry Obligations under International Law" in Marc Bungenberg et al., Eds, *International Investment Law* (Baden-Baden - Munich: H. BECK - Hart – Nomos, 2015), at 685.

[37] *Montenegro Foreign Investment Law*, RM Official Gazette, No. 52/00.

[38] République de Guinée, Assemblée Nationale, Loi L/2015/N°008/AN Portant Code des Investissements de la République de Guinée.

> *La direction effective des entreprises visées à l'alinéa précédent est assurée par des personnes physiques de nationalité guinéenne résidant en Guinée.* [emphasis added].

The relevance of the international investment law concept of local equity requirements with respect to space law can be better understood by a closer examination of the historical setting of this concept. Local equity requirements can be explained by historico-ideological factors. On the one hand, the appearance of such requirements in many developing countries that were once under colonial rule goes hand in hand with the decolonization process of the 1960s. On the other hand, in the states of Eastern Europe, local equity requirements were common prior to the fall of communism as they enabled socialist states to "marry socialist ideology with the admission of foreign investment on the ground that ultimate control over the investment remained with the state."[39] This pattern remained dominant in these states even after the fall of communism and spread to communist Asian states such as China where such local equity requirements remain an important means through which FDI take place.[40]

In both newly independent countries and socialist states, the core elements that led to a broader inclusion of local equity requirements in domestic investment laws are similar. Indeed, willing to reaffirm their sovereignty, newly independent countries nationalized as many sectors of their economies as they possibly could by extending their public sectors and creating state-owned entities (SOEs) to carry a vast array of economic activities under their control, including manufacturing, transportation, insurance, and retail sales. This period marked the apogee of the "fundamental belief that their [i.e., developing countries] governments had the primary responsibility for bringing about economic development"[41] after they became independent,[42] that development could not take place unless the state started planning and directing economic and social life,[43] and that it could not be achieved without veering away from dependence on foreign investment.[44]

---

[39] Sornarajah, *supra* note 28, at 106.

[40] *Catalogue of Industries for Guiding Foreign Investment* (No. 4 Decree of the National Development and Reform Commission and the Ministry of Commerce of the People's Republic of China), online: http://www.fdi.gov.cn/1800000121_39_4851_0_7.html.

[41] International Bank for Reconstruction and Development, *World Development Report 1997 – The State in a Changing World* (Oxford: Oxford University Press, 1997), at 1–2. See also, Salacuse, *supra* note 35, at 60.

[42] See, for example, art 166 (al. 1) of the 1982 Constitution of Turkey (*Constitution of the Republic of Turkey*, 7 November 1982):

> The planning of economic, social and cultural development, in particular the speedy, balanced, and harmonious development of industry and agriculture throughout the country, and the efficient use of national resources on the basis of detailed analysis and assessment and the establishment of the necessary organisation for this purpose are the duty of the State.

[43] Edward S Mason, *Economic Planning in Underdeveloped Areas: Government and Business* (New York: Fordham University Press, 1958), at x ("The plain fact is that throughout the underdeveloped world, the pressures for economic development are all but irresistible. A government that fails to seize the levers of economic development, or at least make the attempt, is probably not long for this world"). See also International Bank for Reconstruction and Development, *supra* note 41, at 1–2.

[44] Besides Marxist theories which do not recognize the concept of property that is key with respect

Decades after the privatization movement replaced the state-planning model, local equity requirements are still provided in many domestic laws. The justification for such requirements under the state-planning model was clear: the state is the main regulator and actor of the economy. Accordingly, when foreign investors are allowed to enter the domestic market, it can only be done after ensuring that the control of all economic operations ultimately rests with the state. This justification is not applicable in the privatization model which is characterized by the disinvolvement of the state from economic affairs. Accordingly, new justifications have been proposed for local equity requirements in a privatized economy.

Two main categories of local equity requirements can be distinguished: the requirement to enter into a joint venture with the state or a state agency and the requirement to enter into a joint venture with a private local partner. With respect to the former, such requirements are nowadays often justified on grounds of national or economic security and the protection of key national interests, i.e., to avoid seeing economic sectors falling under the control of foreigners.[45] Such sectors often include transportation or the exploitation of natural resources, for example.

With respect to local equity requirements mandating foreign investors to penetrate a host state market through a joint venture with a private local partner, they have often been justified on the basis of "indigenization" or "ethnic" policies. Such policies aim at guaranteeing the participation of local populations in the economy of the host state. In Malaysia, for instance, as part of the country's 1970s "New Economic Policy," preference was given to *Bumiputra* (indigenous people) which required specific shareholding arrangements to ensure the participation of locals in the domestic economy.[46]

Additionally, local equity requirements imposing the participation of local capital (whether public or private) and in particular the participation of a local partner have been justified on the grounds that it leads to the maximization of the investment's potential in terms of host state development by allowing for a more efficient transfer of management skills and technology to that host state and the development of domestic human capital[47] and that it allows for a faster and easier penetration of the local market by foreign investors.

---

to international investment law, a wide spectrum of opinions exist in this regard; while some believe in the benefits of FDI, others emphasize the need to avoid relying on such investments. Some approaches are also situated in between these extremes.

[45] See UNCTAD, *Foreign Direct Investment and Performance Requirements: New Evidence from Selected Countries* (United Nations, 2003), at 6–9.

[46] See, for example, R S Milne, "The Politics of Malaysia's New Economic Policy" (1976) 49(2) *Pacific Affairs* 235 and R. Thillainathan and Kee-Cheok Cheong, "Malaysia's New Economic Policy, Growth and Distribution: Revisiting the Debate" (2016) 53(1) *Malaysian Journal of Economic Studies* 51.

[47] Magnus Blomström, Ari Kokko, and Mario Zejan, "Why do Multinationals Seek Out Joint Ventures?" in Magnus Blomström, Ari Kokko and Mario Zejan, Eds, *Foreign Direct Investment: Firm and Host Country Strategies* (Basingstoke: Palgrave Macmillan, 2000), at 30:

> A number of measures intended to encourage multinational firms to transfer more technology have been introduced over the years, including requirements for a certain degree of local participation in the ownership of the MNCs' affiliates. By forcing the multinationals to 'unbundle the package' of inputs they bring to the host country, it is believed that a host country can enable local firms to obtain access to these inputs without the traditional direct investment.

Therefore, it can be argued that the *raison d'être* of local equity requirements is to prevent a locality from losing its identity, its resources, and its potential for future operations. At the same time, such requirements constitute guarantees of a good behavior (or at least of control over the investor's behavior), and, as such, they enhance legal certainty and foreseeability for both the investor and the host state. Similar reasons would justify importing local equity requirements in the realm of investments involving the use of outer space areas.

It is true, however, that the establishment of such a mechanism will not be an easy task as far as the uses of outer space are concerned. At the national and bilateral levels, the inclusion of local equity requirements in domestic legislation and the enforcement of such requirements are a matter of national policy. In this regard, the host state plays and acts as the overarching authority that controls the operation and enforces the requirements. On the contrary, no such overarching authority or institution exists in the governance of outer space. However, the responsibility mechanism that exists in space law is of key importance to fill this gap. In particular, by making states responsible for the undertaking of national public and private space activities in accordance with the international obligations of the state,[48] the latter automatically become bound to ensure that unilateralism in the exploration and use of outer space will not take place as far as their nationals are concerned. Therefore, it would be the state's obligation to impose standards similar to the local equity requirements to its private investors in order to ensure that the activity is undertaken in accordance with the state's international obligations and, thus, in the benefit of humanity in its entirety. By contrast, the adoption of national laws promoting oligopolies in the exploration and use of outer space by attributing exclusive rights over parts of outer space to their nationals[49] can become a burden rather than a blessing for these states as such laws constitute proof of the state's intentions to ab initio act against the standard of global commons that international space law attributes to the uses of outer space.

4.5 The Example of Local Equity Requirements as a Tool Toward Global Access to Space Technology

One more advantage that the use of the mechanism of local equity requirements can bring to the exploration and use of outer space relates to the problem of technology transfer. Space activities, such as the access and exploitation of the natural resources of celestial bodies, require sophisticated technology that is developed and controlled by a small number of public and private space actors.[50] As such, the oligopolistic access, exploration, and use of outer space can easily occur in practice if the space

[48] *Outer Space Treaty*, *supra* note 10, art VI.

[49] See, for example, *US Commercial Space Launch Competitiveness Act* and *Luxembourg Draft Law on the Exploration and Use of Celestial Bodies*, *supra* note 3.

[50] See, generally, Alessandra Vernille, *The Rise of Private Actors in the Space Sector* (Switzerland: Springer, ESPI, 2018).

technology and know-how is not disseminated. A similar problem had also been observed in the case of FDI, and part of the solution was based on the use of local equity requirements.

Indeed, one of the most important problems faced by developing countries, and especially by least developed countries, consists of "[l]ow economic growth rates and low standards of living [that] are endemic."[51] Furthermore, technology has been recognized as being of paramount importance in order to enable developing and least developed countries to reach a state of development similar to that of developed countries: technology underdevelopment is one of the major causes of "basic internal problems such as the inability to feed the local populations or to provide routine health services."[52]

Technology transfer, i.e., "the attempt to fill the gap between countries in current technology (both hardware and production methods) or the ability to discover and innovate,"[53] and technological advancement, i.e., the "primary catalyst towards significant economic development,"[54] are key to allow developing countries that "lack the necessary technological knowledge to catch up with the developed world's standard of living."[55]

In order to enable developing countries to innovate in a way that is adapted to their specific needs without having to pay royalties outside the country, priority should be given to the "use of local or native research and development."[56] However, importing advanced technology from outside into their markets[57] has numerous advantages for developing countries among which the necessity it leads to for the importing country to invest in local infrastructures as well as the training of technicians and specialists in order to exploit the technology.

---

[51] Geoffrey Kransdorf, "Intellectual Property, Trade, and Technology Transfer Law: The United States and Mexico," 7:2 *B.C. Third World L.J.* 277 (1987), at 279.

[52] Helen E. Weidner, "The United States and North-South Technology Transfer: Some Practical and Legal Obstacles," 2 *Wis. Int'l L. J.* 205 (1983), at 205.

[53] William Roscoe Kintner and Harvey Sicherman, *Technology and International Politics: The Crisis of Wishing* (Massachusetts: Lexington Books, 1975), at 91–92.

[54] Geoffrey Kransdorf, *supra* note 51, at 279. See also Michael Blakeney, "Transfer of Technology and Developing Nations," 11:4 *Fordham Int'l L. J.* 689 (1988), at 693.

[55] Helen E Weidner, *supra* note 52, at 205.

[56] Geoffrey Kransdorf, *supra* note 51, at 280. See also President Reagan, "International Meeting on Cooperation and Development – Statement at the First Plenary Session of the Meeting in Cancun, Mexico. October 22, 1981," 17 *Weekly Comp. Pres. Docs.* 1185 (1981), at 1185–1186:

> [g]overnment has an important role in helping develop a country's economic foundation. But the critical test is whether government is genuinely working to liberate individuals by creating incentives to work, save, invest, and succeed. ... Because whenever they are allowed to create and build, wherever they are given a personal stake in deciding economic policies and benefiting from their success, then societies become more dynamic, prosperous, progressive, and free.

[57] Gerald J Mossinghoff, "The Importance of Intellectual Property Protection in International Trade," 7 *B.C. Int'l & Comp. L. Rev.* 235 (1984), at 245.

In this regard, FDI is one of the most important channels through which transfers of technology occur in host state. However, it has been observed that "[t]he traditional method of transferring technology has been through investment in wholly owned and controlled subsidiaries."[58] The disadvantage with such structures is that it allows the foreign investors to keep their technology protected and limits its dissemination in the host state. By contrast, local equity requirements and the partnership that they require have the advantage of both bringing technology in the host state and, via the collaboration with the local partner, allowing wider dissemination in that state.

A similar use of local requirements for technology transfers in the domain of outer space investments could prove to be particularly beneficial and allow for the dissemination of advanced technology that currently belongs to a small group of public and private actors. However, the use of such requirements and their spillover effect of disseminating technology raise one main question: What would make the investor abide by such requirements leading to transfers of technology and not take the road of de facto unilateral use of outer space given that it has the advantageous position of possessing the necessary technology. The answer to this question can, again, be located in the responsibility regime of space law, which would require that the authorization of the activity and its legal undertaking demands some sort of partnership with public and private entities of other states and the subsequent technology and knowledge transfer.

Such cooperation in technology and knowledge exchanges is not only required by the letter and spirit of the Outer Space Treaty. It is also critical as a matter of necessity as space activities most often require the combination of space technologies, not all of which are owned by or accessible to one isolated actor. Therefore, cooperation in the field of space technologies is necessary for the advancement of investments with the use of outer space as their object. In this regard, the International Space Station stands as a good example of such cooperation that has been successful.[59] Subsequently, even if investors that possess the know-how do not wish to disseminate it, standards could be adopted to activate access to it and not necessarily allow for the unconditioned transfer of it. A mechanism equivalent to local requirements for technology access and transfer could constitute a guiding example.

---

[58] David M. Haug, "The International Transfer of Technology: Lessons that East Europe Can Learn from the Failed Third World Experience," 5 *Harvard Journal of Law & Technology* 209 (1992), at 213.

[59] Michael Mineiro, *Space Technology Export Controls and International Cooperation in Outer Space* (Switzerland: Springer, 2012), at 161.

4.6 Conclusion

Cooperation is necessary in any kind of activity that involves a cross-border element, be it at the bilateral, multilateral, international, or global levels. It is increasingly difficult for a single actor to own all the necessary resources to successfully undertake complex operations: some actors possess financial resources, while others own the necessary technology and know-how. In addition, collaboration can also be critical to undertake cross-jurisdiction activities or, in the case of space, activities that take place in a context characterized by the absence of sovereignty. As such, mutual concessions and understanding among the parties involved constitute key factors for the success of such activities regardless of whether two or more states are involved. The crux of the issue that this paper addressed is how to enforce such cooperation in the case of global commons and in particular of outer space.

This contribution illustrates how local equity requirements – and most importantly their rationale – constitute a tool that can efficiently allow taking into account the interests of the locality when foreign actors undertake operations in outer space. To enhance the access of local populations to resources, local equity requirements allow maintaining some control over the behavior of the foreign operator of the investment. In that way local equity requirements applied in the context of investments in outer space could allow to ensure that the key interests and rights of humanity and humankind in space law are preserved.

Chapter 5
From Little Things, Big Things Grow: How Should We Regulate the Commercial Utilization of Small Satellite Technology?

Steven Freeland

5.1 The Changing Nature of Space Technology

October 1957 witnessed the launch of the first human-made space object to orbit the Earth, Sputnik 1. Since that time, there has been a breathtaking and seemingly endless development of space-related technology. Humankind is now engaged in a multitude of space activities far beyond the contemplation of those involved at that time. The utilization of space technology now forms a crucial part of everyday society in all parts of the globe – irrespective of the (geo)political, economic, and cultural characteristics of any one country. Simply put, our reliance on space technology is such that the world would cease to function in many respects without constant and unimpeded access, and this imperative is likely to become even more pronounced for future generations. This has primarily been driven by the increasing "commercialization" of outer space.

Yet, as is well known, there remains a vast gulf between the space capabilities of the relatively small number of space "powers" compared with the rest of the world.

This chapter was written in September 2015.

S. Freeland (✉)
Western Sydney University, Sydney, NSW, Australia

iCourts Centre of Excellence for International Courts,
University of Copenhagen, Copenhagen, Denmark

University of Vienna, Vienna, Austria

International Institute of Space Law, Paris, France

Space Law Committee, International Law Association, London, UK

London Institute of Space Policy and Law, London, UK

European Centre for Space Law, Paris, France

Australian Centre for Space Engineering Research, Kensington, NSW, Australia
e-mail: S.Freeland@westernsydney.edu.au

G. D. Kyriakopoulos, M. Manoli (eds.), *The Space Treaties at Crossroads*,
https://doi.org/10.1007/978-3-030-01479-7_5

It has been estimated that approximately up to 60 States now have some form of direct space capability, although the extent that they are able to utilize space for their own development (and other) purposes varies quite significantly. Of course, this also means that perhaps up to 140 States thus far have not realistically *any* independent capability to directly access space themselves, this despite their reliance on the technology for many aspects of their functioning and development. These countries are instead totally dependent on others for their space access, which therefore impacts upon their space "security" and impedes opportunities for creativity, innovation, and progress among their citizens. The reality is that their access to satellite data and the ability to utilize vital space technology in a crisis would be largely dependent on and subject to the strength and enforceability of their existing contractual relationships and political ties.

It is in this context that the recent development and adaptation of so-called "small" satellite technology potentially represent a paradigm shift in the way humankind accesses space. These satellites are usually cheaper and less complex to develop, build, and launch than conventional satellites. They therefore open the possibilities for a significantly greater degree of space access to a much larger range of space "actors." Already, groups such as university students and nonprofit organizations in both developed and developing countries have increasingly been able to become involved in space through these means. The development of this technology may represent an important precursor to the establishment of indigenous and independent space programs in States that previously could not have considered such activities. In effect, by eliminating some significant barriers to entry, small satellite technology may facilitate capacity building, broader collaborative opportunities, and education/training programs, as well as bridging (some) technology gaps, for hitherto "non-spacefaring" States. It will also open up even more diverse commercial opportunities for a much broader range of potential service providers and, generally, "bring space to more people."

Significantly, also, as the technology develops even further, it may also open the door to "traditional" users of outer space – both States but also private commercial entities – to utilize it for existing as well as new purposes, thus expanding the scope of their capability at a significantly lower relative cost. Of course, this may require a mind-shift on the part of existing space actors, as they grapple with whether, and how to adapt to this relatively new technology and adjust their activities to react to the challenges posed by the potential for new market entrants.

As a consequence, therefore, the increasing advent of this technology could potentially redefine the landscape of many activities in space. This new space paradigm will not see the end of more traditional satellite technology since, naturally, small satellite technology will not quench our insatiable demand for all that space can provide. However, it does open up a plethora of possibilities, many of which we are simply not in a position to comprehend or even imagine at this point. In this regard, one might liken the potential of small satellites to the way that mobile phones have revolutionized terrestrial communications activities. We simply do not know where this technology might ultimately lead and what it will allow us to do. However, we can confidently expect that it will open the door to an even more expansive array of commercial opportunities.

Thus, from a technological perspective at least, small satellite technology most likely represents a "win-win" possibility that enhances the momentum for change and further promotes commercial space activities. Indeed, in many respects, this has been the singular motivation for both developers and users thus far. As with many aspects related to the exploration and use of outer space, the technology continues to move forward at a rapid pace without sufficient attention being paid to the regulatory consequences and requirements. It is therefore important not to be too caught up in this wave of optimism and innovation, without at least also considering how these developments coexist with the current regulatory framework, which has largely been designed with "big" satellite technology in mind.

The purpose of this paper is therefore to take pause and reflect on various regulatory requirements and challenges posed by the existing international legal regime in relation to the use of small satellite technology. While many of the users of this technology are no doubt cognizant of these requirements, it is probably fair to say that many are not; or, put another way, they do not consider the regulatory issues with the same degree of attention as they do the technical factors.

What this discussion will highlight is the fact that the existing legal framework was not designed with small satellite technology specifically in mind. As a result, further regulation will be required – particularly at the national level – and this will necessitate a "balancing" of sometimes competing interests between protecting the State now and into the future from potentially very significant liability on the one hand and encouraging innovation and research and development on the other. Although the discussion below focuses on the current regulatory requirements, it leads to the conclusion that the design of future legal regimes to deal specifically with small satellite technology will necessitate some fundamental policy decisions by national lawmakers and regulatory bodies.

5.2 The Current International Legal Framework and Regulatory Requirements

The international regulation of the exploration and use of outer space is primarily based upon a series of five United Nations Space Treaties[1] and several General Assembly Principles.[2] The Treaties in particular set out a number of fundamental

[1] These are (i) 1967 Treaty on Principles Governing the Activities of States in the Exploration and Use of Outer Space, including the Moon and other Celestial Bodies, 610 U.N.T.S. 205 (Outer Space Treaty); (ii) 1968 Agreement on the Rescue of Astronauts, the Return of Astronauts and the Return of Objects Launched into Outer Space, 672 U.N.T.S. 119 (Rescue Agreement); (iii) 1972 Convention on International Liability for Damage Caused by Space Objects, 961 U.N.T.S. 187 (Liability Convention); (iv) 1975 Convention on Registration of Objects Launched into Outer Space, 1023 U.N.T.S. 15 (Registration Agreement); and (v) 1979 Agreement Governing the Activities of States on the Moon and other Celestial Bodies, 1363 U.N.T.S 3 (Moon Agreement).

[2] See, in particular, (i) 1963 Declaration of Legal Principles Governing the Activities of States in the Exploration and Use of Outer Space, United Nations General Assembly Resolution No 1962;

rules, imposing various obligations on States Parties, some of which are also regarded as representing customary international law.[3] More and more States have come to recognize the need to promulgate national space laws to "transform" these international obligations into their respective domestic legal spheres.[4] Given that the advent of small satellite technology presents opportunities for hitherto non-spacefaring States to engage in space activities, it may well be that the development of such technology in a particular country may "pre-date" any specific applicable national laws. Thus, the possibilities of greater access to this technology may be a driving force in the enactment of a further "wave" of national space law in various countries – for example, as was the case in Austria, which enacted its national space law in late 2011.

It should be noted that, in addition to these various instruments, there have recently been an increasing number of "soft-law" guidelines concluded that also relate to the conduct of particular activities in outer space. This has been for several reasons, partly related to the strategic and political nature of space, which has made the finalization of internationally binding treaties more difficult to achieve.[5]

This paper will refer primarily to existing "hard law" regulatory requirements that flow from the Space Treaties – although reference will be made to one important set of voluntary guidelines – from the perspective of how they may relate to the use of small satellites, and seek to raise some pertinent questions that arise from their applicability. It is not intended in this paper to be exhaustive in this regard, or comprehensive as to all precise details, but rather to raise the more significant issues and the challenges they pose. This will also serve to highlight the importance of properly addressing this issue by way of specifically directed regulation, given that the use of small satellite technology will most likely continue to grow exponentially into the future.

(ii) 1982 Principles Governing the Use by States of Artificial Earth Satellites for International Direct Television Broadcasting, United Nations General Assembly Resolution No 37/92; (iii) 1986 Principles Relating to Remote Sensing of the Earth from Outer Space, United Nations General Assembly Resolution No 41/65; (iv) 1992 Principles Relevant to the Use of Nuclear Power Sources in Outer Space, United Nations General Assembly Resolution No 47/68; and (v) 1996 Declaration on International Cooperation in the Exploration and Use of Outer Space for the Benefit and in the Interest of All States, Taking into Particular Account the Needs of Developing Countries, United Nations General Assembly Resolution No 51/122.

[3] See, generally, Stephan Hobe, Bernhard Schmidt-Tedd and Kai-Uwe Schrogl (eds), *Cologne Commentary on Space Law, Volume I – Outer Space Treaty* (2009), and *Volume II – Rescue Agreement, Liability Convention, Registration Convention, Moon Agreement* (2013).

[4] See United Nations Office of Outer Space Affairs (UNOOSA) website, "National Space Law Database" http://www.unoosa.org/oosa/en/SpaceLaw/national/state-index.html.

[5] For a discussion of the increasing trend toward the conclusion of non-binding instruments in the realm of outer space, see Irmgard Marboe (ed), *Soft Law in Outer Space: The Function of Non-binding Norms in International Space Law* (2012); Steven Freeland, "For Better or For Worse? The Use of "Soft Law" within the International Legal Regulation of Outer Space" (2011) XXXVI *Annals of Air and Space Law* 409.

Some of these regulatory requirements are as follows.

(i) International Responsibility: Authorization and Supervision

The regime for space activities is structured on the basis that States bear international responsibility for "national activities in outer space," including when such activities are carried on by non-governmental entities.[6] While there is no precise definition in the Outer Space Treaty as to what constitutes a "national" activity, the terms of the domestic space law of a particular State will clarify the scope of activities to which it refers – in essence, representing an interpretation by the drafters of that legislation of what they regard to be national activities in outer space, at least for the purposes of the specific domestic law.

A review of existing national space law indicates that, in most cases, States have legislated for the regulation of space activities based on the "territoriality" of the activity (i.e., where an activity, e.g., a launch, involves the territory of that State), in accordance with general international law principles of jurisdiction. In addition, many States that have national space law also regulate space activities based on the nationality of the space actor (i.e., the person/entity engaged in the space activity). For example, the Australian *Space Activities Act* (1998)[7] provides that certain space activities carried out in Australia, or by an Australian national outside Australia, are subject to regulation under the legislation and require an appropriate approval under the licensing system it establishes. Thus, a launch of a small satellite in Australia by an Australian University will engage the international responsibility of that State under the Outer Space Treaty. Likewise, so will the involvement of that University in a small satellite program – for example, the QB50 program[8] – where the satellites are to be launched from another State. In these circumstances, therefore, (international) responsibility under the Outer Space Treaty is interpreted as extending to extraterritorial activities.

Article VI of the Outer Space Treaty goes on to require that the "appropriate State" – which is thought by most commentators to mean the State whose national activity it is – undertakes the "authorization and continuing supervision" of such activities. Typically, the authorization of space activities is implemented by way of a licensing regime established under national law (at least for those States with specific domestic space legislation).[9] This can be through the creation of a com-

[6] Outer Space Treaty, article VI.

[7] *Space Activities Act*, Australia, No. 123, 1998.

[8] The QB50 mission involves the launching in 2015 of a network of 50 'CubeSats' built by Universities all over the world as a primary payload, with the aim of performing various scientific experiments in the lower thermosphere at an altitude of approximately 320 kilometers: see "QB50, an FP7 Project: Mission Objectives," https://www.qb50.eu/index.php/project-description-obj. Australian Universities are involved in this project: Australian Centre for Space Engineering Research, "QB50 – an International Network of 50 CubeSats" http://www.acser.unsw.edu.au/QB50/index.html.

[9] See Steven Freeland, "Matching Detail with Practice: The Essential Elements of National Space Legislation" (2010) *Proceedings of the International Institute of Space Law*, 540.

prehensive "one-size-fits-all" license regime or, more likely, via the establishment of different forms of license, depending upon the particular space activity for which authorization is being sought. For example, the Australian *Space Activities Act* creates a number of different licenses to deal with specific space (launch-related) activities, including a "Launch Permit" for launches from Australian territory and an "Overseas Launch Certificate" for launches of a space object by an Australian national from launch facilities outside of Australia.

In relation to the use of small satellites, there is little conjecture that their launch and use do, indeed, constitute a space activity. Moreover, the satellite itself would in most circumstances be a space object for the purposes of international space law – including for the purposes of the Liability Convention (see below), as well as the domestic law of most countries. Activities involving small satellites therefore would typically fall within the scope of article VI of the Outer Space Treaty. This in itself is not surprising – what is, however, is that this is not necessarily understood by the users of small satellites, particularly with respect to "experimental" projects. The reality is that those seeking to engage in small satellite activities, irrespective of where those satellites might be launched, should take careful note of the relevant national laws and apply for the requisite license (where applicable). As noted below, this might also have added consequences in terms of financial and liability concerns, as well as other aspects of conditionality.

Moreover, the requirement of continuing supervision on the part of the State may be quite complex. There is, for example, some conjecture as to how, in practice, the need for continuing supervision might be undertaken in circumstances where the relevant space activity is a cooperative venture between institutions in a number of States. Internal arrangements between the cooperating States should be put into place to allow for each State to, in some way, exercise a degree of supervision, at least in relation to those aspects of the activity (and over its nationals who may be involved in its ongoing operation) in which it has a specific interest. Yet, even this presupposes that the institutions or persons engaged in the small satellite activity have informed the relevant governmental agency of their involvement and have provided specific details as to the scope of the program, design, issues of control, etc.

Adding to the complexity is the fact that most small satellites are not designed with control systems and therefore cannot be maneuvered once they are launched and operative. As soon as they are placed in orbit, their position cannot be altered from Earth. This may also explain why this requirement may often have been disregarded, leaving the responsible State in a difficult position in terms of its obligations under the Outer Space Treaty.

(ii) International Liability: National Indemnity Requirements

The general international liability provisions found in the Outer Space Treaty[10] and the more detailed regime specified in the Liability Convention impose liability

[10] Article VII of the Outer Space Treaty prescribes the general terms giving rise to international liability for damage caused by an object launched into outer space. The scope of international liability is then elaborated in the Liability Convention. However, even if it is not a State Party to the

on a "launching State" for certain damage caused by a space object.[11] There are no time limitations or caps on the amount of this liability under the Liability Convention, as long as it represents "damage"[12] by a "space object" as those terms are defined for the purposes of that Treaty.[13] In the absence of specific indemnities in relation to claims by third parties, or where the various exceptions and exonerations contained in the Liability Convention do not apply, a launching State will bear this international obligation of liability[14] even in circumstances where the space activity is undertaken by a non-governmental entity and perhaps also even where the State may not be aware of the activity at all.[15]

This represents one compelling "incentive" for States to pass domestic space law. The enactment of national space law enables States to formalize domestic legal processes that would allow them to pass on financial responsibility to and recover from their national non-governmental entities the full amount (or part thereof) of the damages for which the State may be liable at the international level. Of course, this does not remove the international obligation of liability of a launching State under the Liability Convention – this contingent liability remains in place in relation to any space object for which a particular State is deemed to be a launching State. However, it does enable the State to put in place a domestic mechanism by which it can transfer the financial "risk" associated with this potential international liability for third party claims. Indeed, this is precisely what a number of States have done in their national laws in relation to "traditional" satellite technology. For example, one of the objectives of the Australian *Space Activities Act* is "to provide for the payment of adequate compensation for damage caused to persons or property as a result of space activities regulated by [the legislation]."[16]

Liability Convention, a State would still be subject to the liability provisions in the Outer Space Treaty, as well as any other potential claims based on the general public international law principles of State responsibility.

[11] The identity of the relevant launching State(s) is determined at the time of launch, with article I(c) of the Liability Convention defining a launching State as:

(i) "A State which launches or procures the launching of a space object;
(ii) A State from whose territory or facility a space object is launched"

[12] Article I(a) of the Liability Convention defines "damage" as:

> "… loss of life, personal injury or other impairment of health; or loss of or damage to property of States or of persons, natural or juridical, or property of international intergovernmental organizations."

[13] As noted, it would be difficult to argue that an operating small satellite was *not* a space object for the purposes of the Liability Convention, even if it is not maneuverable while in operation.

[14] See Steven Freeland, "There's a Satellite in my Backyard! – Mir and the Convention on International Liability For Damage Caused by Space Objects," (2001) 24:2 *University of New South Wales Law Journal* 462.

[15] On this point, there may be an argument that, where the only possible relevant mode by which a State could be a launching State in a specific case is by "procuring" the launch, there is a minimum "threshold" test to demonstrate such procuring, at least based on knowledge of the particular activity. However, it is unclear whether such an argument reflects the correct legal position.

[16] Australian *Space Activities Act*, section 3(b).

As a consequence, national space legislation often attaches conditionality to the issue of a license to engage in a specific space activity, the practical effect of which is to require the applicant to provide or somehow procure an indemnity to the Government for damage, although the amount may be subject to specific caps under particular national law. Although it would be relatively straightforward to simply require the applicant in these circumstances to take out appropriate commercial insurance against third party claims to the extent of the specified (maximum) damage, this would often be impractical (given the relative lack of depth of the international space insurance market) and, more specifically in the case of many small satellite operators, disproportionally costly.[17] Indeed, such a requirement might make the planned small satellite activity unaffordable, thus preventing it from going ahead at all.

This gives rise to difficult considerations that would require a "balancing" between the protection of the State from potential financial liability and the desirability of encouraging expertise, research, and development, perhaps as a precursor to more profitable and commercial opportunities down the track. Such potentially conflicting interests between a need for regulation on the one hand and the provision of incentives for new innovation on the other are not unique to the situation of small satellite operators – similar arguments have been raised in relation to the requirement for the "equitable sharing of benefits" derived from the exploitation of natural resources under the Moon Agreement. However, unlike the Moon Agreement, virtually every spacefaring State is a party to both the Outer Space Treaty and the Liability Convention – and, in any event, the liability regime they establish arguably also reflects customary international law. It is therefore incumbent on all States with an (potential) involvement in space to somehow address this issue.

The ideal scenario would be for the small satellite operator to negotiate with the relevant launch service provider for the provision of insurance cover and/or an indemnity by that provider (and perhaps also the Government standing behind that provider) to the launching State and the payload owner (e.g., the University that has built the small satellite(s)), at least in relation to certain elements of potential third party claims (again most likely subject to a cap). This is often the case in commercial launch service contract arrangements for large satellites. Some small satellite operators contend that the position is more complicated in the case of a collaborative small satellite program such as the QB50 project, involving Universities from several countries (and thus potentially a considerable number of launching States). However, the point remains that many such programs have proceeded without the issue even being raised with either the launch service provider or the intermediary arranging the launch.

Once again, this is something that should be negotiated coincidentally with the development of the technical aspects of such a program. A failure to do so potentially not only places the launching State in a difficult position but might also expose

[17] The Australian *Space Activities Act* envisages that, in certain circumstances, rather than procure insurance, an applicant could instead demonstrate "direct financial responsibility" as an alternative (section 47).

the institution supporting the small satellite operators to a real and unacceptable risk of liability. Obviously, this should be of practical concern to those involved.

(iii) Registration: National and United Nations Registers

The Registration Convention creates a two-pronged regime of registers that are relevant in respect to space objects that are launched inter alia "into earth orbit."[18] The State of Registry (as defined) is to maintain a national register in which such space objects are to be included and, in addition, shall provide certain specified information in relation to those objects to the United Nations, which itself maintains a central register.[19] In accordance with the terms of the Outer Space Treaty, the registration of a space object within a State's national register also has implications with regard to the "jurisdiction and control" of that object.[20]

In situations where a State has not, for example, previously been involved in launching activities, it may not have in place a national register nor a mechanism for the furnishing of the required information to the United Nations. There may be a time lag associated with the establishment of the national register, which, in most circumstances, could only be implemented under national space legislation.[21] Once again, this will require consultation and information flows between the small satellite operator and the relevant Government agency (if indeed such an agency exists).

In addition, with widespread cooperative small satellite programs that may potentially involve institutions from many countries, there will be need for careful coordination between the various launching States as to who should be the State of Registry – it can only be one of the launching States.[22] It may not, for example, make practical sense that each launching State would seek to be the State of Registry for its specific small satellites in the context of a joint program involving a large constellation of objects launched simultaneously from the one launch vehicle.

(iv) Sustainability of the Space Environment: Space Debris Mitigation

One of the major challenges for the future exploration and use of outer space is the growing proliferation of space debris. Much has been written about the exponential growth of pollution in outer space and the hazards that it poses. There are many views as to how the problems should be addressed, given that the whole issue of the environment of outer space is a complex one, with many interconnecting variables at play.[23] As noted above, these variables, and the enormous financial implications that would arise from setting in motion binding requirements, have meant that, to date, only soft-law guidelines, rather than hard law treaty regulation,

---

[18] Registration Convention, article II(1).

[19] *Ibid*, article IV(1).

[20] Outer Space Treaty, article VIII.

[21] See Australian *Space Activities Act*, Part 5.

[22] Registration Convention, article I(c).

[23] See, for example, Ulrike M Bohlmann and Steven Freeland, "The Regulation of Space Activities and the Space Environment" in Shawkat Alam, Md Jahid Hossain Bhuiyan, Tareq M.R. Chowdhury and Erika J. Techera (eds), *Routledge Handbook of International Environmental Law* (2013), 375.

have been agreed to address this issue. Nonetheless, these guidelines,[24] although voluntary and expressed in general terms, are significant in that they reflect the existing practices as developed by a number of States and international organizations and set (minimum) standards toward which spacefaring nations should strive.

The principles underpinning the debris mitigation guidelines are that care should be taken to minimize the risk of debris "creation" in the conduct of space activities.[25] The importance of space for all aspects of our lives necessitates a diligent adherence to these standards to the greatest extent possible. It is generally recognized that it is in the interests of all spacefaring States to follow these guidelines and this is, as noted, increasingly reflected in their practices. The long-term sustainability of outer space activities is a matter of interest and importance for the international community as a whole and is now one of the principal focal points for UNCOPUOS.[26]

There are some potentially significant environmental challenges that arise from the use of small satellite technology. Growing demand and the expanding range of functions and, ultimately, commercial services points to rapid increases in the numbers of small satellites that will be placed into Earth orbit. In order to utilize this technology to achieve "global" coverage, very large constellations of small satellites will be required and are being planned.[27] While these satellites will primarily be placed into a low Earth orbit, projects such as these will "populate" important orbits with a significant number of space objects and increasingly pose a potential collision risk.

Even with respect to the current low-cost small satellite programs, the issue still remains. Many experimental satellite programs have been exactly that – "experimental." They have often utilized existing "off-the-shelf" components, and the

[24] See IADC Space Debris Mitigation Guidelines prepared by the Inter-Agency Space Debris Coordination Committee in 2007 (IADC Debris Mitigation Guidelines) http://www.iadc-online.org/index.cgi?item=docs_pub: and United Nations General Assembly Resolution 62/217, which (in paragraph 26) endorsed the Space Debris Mitigation Guidelines agreed by the United Nations Committee on the Peaceful Uses of Outer Space (UNCOPUOS) (UN Guidelines).

[25] The UN Guidelines recognize two broad categories of space debris mitigation measures: those that curtail the generation of potentially harmful space debris in the near term – minimization of the production of mission-related space debris and the avoidance of break-ups; and those that limit their generation over the longer term - end-of-life procedures that remove decommissioned spacecraft and launch vehicle orbital stages from regions populated by operational spacecraft.

[26] See, for example, United Nations Information Service, Press Release, "Long-term sustainability of outer space activities among the key topics of Scientific and Technical Subcommittee at its 51st Session," UNIS/OS/432, 7 February 2014 http://www.unis.unvienna.org/unis/en/pressrels/2014/unisos432.html.

[27] See, for example, Ellie Zolfagharifard and Sarah Griffiths, "Elon Musk's new mission revealed: SpaceX founder confirms plans for tiny satellites that will provide cheap internet worldwide" *Daily Mail Online*, 12 November 2014 http://www.dailymail.co.uk/sciencetech/article-2830263/Elon-Musk-s-new-mission-revealed-SpaceX-founder-confirms-plans-tiny-satellites-provide-cheap-internet-worldwide.html. This article reports that SpaceX plans to launch 700 satellites (each weighing 113 kg) and Google 180 satellites, both in an effort to provide internet services for the 4.8 billion people of the world who are still without online access.

expectations of mission success for any significant period of time have not necessarily been high. It is fair to say that such circumstances give rise to lower perceptions of risk and a higher tolerance toward failure. For many such programs, at least in the relatively early phases of small satellite development, the process has largely been about the journey (to space) rather than delivery of services – though of course this is now changing. Many of these programs have relied on "piggy-back" launches, which have meant that the satellites have been placed in orbits significantly higher than the very low orbits that would allow them to decay relatively quickly. For many small satellites, therefore, there is a potentially very long period (perhaps in excess of the 25 years cap suggested by the IADC Debris Mitigation Guidelines) before orbital decay, even though the satellite itself will have been functioning for only a short timeframe.

Moreover, as is well known, there are several variants of small satellite technology. While it is too simplistic to categorize them solely on the basis of their size and weight, the so-called "pico" (0.1–1 kg) and "femto" (less than 100 g) satellites may be too small to be picked up by conventional tracking systems. Yet, even such low-mass objects can cause catastrophic damage in certain circumstances. The potential consequences, and therefore the potential risks, would, of course, be greatly magnified should the development of a large-scale commercial human spaceflight industry, despite recent setbacks, ultimately come to fruition.[28]

Of course, these issues are relevant to the question of potential liability raised above. They also point to the need to carefully consider how, and to what extent, the future implementation of small satellite programs can and will be undertaken, so as to as much as possible be consistent with the overarching goal of managing the long-term sustainability of outer space activities in such a way as to maximize the (commercial) benefits that can be derived, while maintaining appropriate and acceptable safety standards, particularly for missions involving humans.

In some senses, therefore, the environmental consequences relating to small satellite programs have not really been properly factored into the regulatory framework. This is also a question of education and awareness but is a highly important factor to take into account when designing the future legal regime to apply to such programs.

(v) Other Regulatory Considerations: Frequency Allocation and Traffic Management

As noted, these brief comments do not purport to be comprehensive as to the relevant regulatory factors associated with this new commercial space paradigm featuring large-scale small satellite programs. However, the primary regulatory

[28] See, for example, Andrea Peterson, "Manned commercial spaceflight: The final unregulated frontier" *The Washington Post*, 6 November 2014 http://www.washingtonpost.com/blogs/the-switch/wp/2014/11/06/manned-commercial-space-flight-the-final-unregulated-frontier/. For a discussion of the relevant legal issues related to the proposals to establish a commercial human spaceflight industry, see Steven Freeland, "Fly Me to the Moon: How Will International Law Cope with Commercial Space Tourism?" (2010) 11:1 *Melbourne Journal of International Law* 90.

issues that ultimately stem from the principal requirements under the United Nations Space Treaties have been raised. There are, of course, other equally relevant considerations that also arise. For example, as more such programs emerge, particularly offering commercial services, the issue of radio-frequency usage becomes all important. The existing use of the "amateur band" frequencies will no longer be applicable and appropriate. The regulatory framework of the International Telecommunications Union (ITU) will become even more relevant. While the ITU operates effectively to manage the use of radio spectra, it is highly bureaucratic. Decisions about allocations of valuable (commercial) frequencies take significant periods of time and are sometimes highly political. The coordination of frequencies so as to minimize harmful interference is complex. This lengthy process does not necessarily sit comfortably with the shorter timeframes associated with small satellite programs, and procedures will need to be established to accommodate this technology without compromising the important work of the ITU. This will not be an easy task.

In addition, the introduction of large numbers of small satellites will highlight even more the imperatives to consider the development of international traffic management systems involving space traffic, as well as its intersection with air traffic. Once again, while some initial steps are being taken to consider these issues,[29] there is much work to be done by all stakeholders.

5.3 Concluding Remarks

This paper has highlighted the fact that the current international legal framework continues to apply to new and developing technologies – such as small satellites – that will contribute to the further evolution of commercial space activities. Of course, the business case for those large programs that have been announced is yet to be proven, and, while it is clear that small satellites will form a (significant) part of the future dimension of space, there may be some "false starts" along the way as to the most appropriate approach to be undertaken by those entities seeking to utilize the technology to develop very significant commercial businesses.

As a consequence, the existing law and the technology, at least at the international level, do not represent a natural fit. The international regulatory framework was not designed specifically to deal with the advent of this technology, nor for the expansive range of new space actors. Moreover, these new actors in particular may not be completely aware of, or understand, the relevance and implications of the existing framework.

The United Nations is therefore conscious of the imperative to explore the potential dynamics of the small satellite industry and promote the need to address both the challenges and the opportunities posed by small satellites. It has, for example,

[29] For example, from 18–20 March 2015, the United Nations Office of Outer Space Affairs (UNOOSA) and the International Civil Aviation Organization (ICAO) jointly sponsored an "AeroSPACE" symposium where some of these issues were discussed.

established the Basic Space Technology Initiative (BSTI), which seeks to support capacity building in fundamental space technology and also to promote the use of space technology and its applications for sustainable development. This has partially been guided by the growth of small satellites technology and the increasing access to this technology by universities and smaller institutions, in countries along the full spectrum of economic development.[30] The BSTI represents a useful international cooperative attempt to better understand the dimensions of the issues that will arise. Despite these initiatives, however, it seems unlikely that binding international frameworks will be put in place anytime soon to effectively deal with this technology.

Yet, even putting these initiatives aside, it is clear that such shifts in space technology require the development of appropriate regulatory standards in a relatively short timeframe. Small satellite entrepreneurs are anxious that any real (or perceived) barriers to entry posed by national regulatory requirements are removed. Many of these entities believe that if they are not able to develop and implement their plans in the short term, then the opportunity will be lost, since someone else will do it instead, perhaps in a more "user-friendly" domestic regulatory environment.

Whether or not these fears are justified in every case, what seems increasingly likely is that, in some respects, small satellite technology will become one mainstream means of utilizing space for commercial purposes in the future. Attempting to regulate this twenty-first-century technology solely by reference to twentieth-century rules is therefore likely to create difficulties and uncertainties and perhaps deter some who would otherwise consider engaging in the space industry.

In the meantime, however, there is no doubt that small satellite technology can offer great opportunities, but it also poses some significant challenges to the broader perspective of the exploration and use of outer space. The need for clear regulation to specifically address this technology is clear, and it thus falls on national lawmakers to provide what is required within a more expedient timeframe. Pressure is already being exerted by industry associations and representatives in various States seeking clarification of the regulatory requirements in relation to this new technology.[31]

In the end, therefore, clear domestic policies must be formulated. National legislatures have to come to grips with the ever-changing range of space technology, particularly if they wish to become increasingly involved in space activities. Some Governments are already attempting through their legislation to deal specifically with the issues that arise through the advent of small satellite technology,[32] but there

[30] See UNOOSA, "Basic Space Technology Initiative (BSTI)" http://www.unoosa.org/oosa/en/SAP/bsti/index.html.

[31] See, for example, Space Industry Association of Australia, "Discussion paper on the Regulatory Requirements for Launches of Small Satellites for Scientific and Educational Purposes", February 2015 (copy with author).

[32] See, for example, Irmgard Marboe and Karin Traunmuller, "Small Satellites and Small States: New Incentives for National Space Legislation" (2012) *Journal of Space Law* 289, where the authors describe how the national laws of Austria, Belgium and The Netherlands have been structured to deal with the possibility of future small satellite programs involving those countries.

is a long way to go. Whatever rules are put in place must find the right balance between, on the one hand, the need for regulation of the financial and technical elements, so as to minimize the risks to an acceptable level, and the facilitation of research and innovation to allow for greater and more efficient access to space and the potential for commercial returns, on the other.

Public policy questions arise as to whether, for example, to exempt small satellite operators from several of the existing national regulatory requirements that apply to their large satellite "brethren." Yet, to do so may have the ultimate effect of minimizing the "incentives" or motivation of these operators to engage in best practices or to take simple, inexpensive steps to ensure that their local stakeholders are covered by existing protections. Naturally, this may not necessarily be the case when it comes to commercial small satellite enterprises; however, it is suggested that the industry as a whole would not necessarily be unduly stifled by the requirement that, in all circumstances, they take proper and appropriate risk management steps. Any relaxation of the rules for the users of this technology will bring with it added risks for the regulators and the relevant State, even though in many cases these might be quite small.

These are difficult choices and States will take differing paths, depending upon their specific circumstances. This will, unfortunately, mean that there is unlikely to be established a uniform international set of rules to address the complexities of small satellites, at least in the short-medium term. Perhaps we might see the emergence of a soft-law code of conduct at the international level, but this may not provide a sufficient base to determine the conduct of those new actors in the space paradigm.

This again points to the strong role that national law and lawmakers have to play, which will require close consultation between all stakeholders and emphasize the need for regulators, the scientific community, the entrepreneurs, and the lawyers to all talk to each other to a far greater degree than has thus far been the case.

Chapter 6
Using Space Objects in Orbit as Transaction Objects: Issues of Liability and Registration *de lege lata* and *de lege ferenda*

Michael Chatzipanagiotis

In modern economy, the fact that some objects are moving in outer space does not deprive them of their fundamental attribute: they are objects, which have a monetary value as long as they can be employed for profit. As a result, such value can be used and is indeed widely used in transactions.

At the same time, space objects are governed by international space law, which was drafted about five decades ago, when space objects served mainly purposes of political prestige.

This dissonance between current uses of space objects and underlying rationale of international law leads to practical problems, related mainly to their registration, which is connected with a variety of consequences and liability. We shall analyse these problems and propose solutions at the example of air law, adjusted to the particularities of space operations.

6.1 Registration of Space Objects under Current International Law

International rules on registration of space objects can be found in the Convention on Registration of Objects Launched into Outer Space (Registration Convention [RC])[1] and the UN General Assembly Resolution 1721B (XVI) on International Co-operation in the peaceful use of outer space [GA Resolution 1721B].[2]

---

[1] *Convention on Registration of Objects Launched into Outer Space*, 6 June 1975, 28 UST 695, 1023 UNTS 15 (entered into force 15 September 1976) [Registration Convention].

[2] Resolution 1721B (XVI) International Co-operation in the peaceful use of outer space, 1085th plenary meeting, 20 December 1961, online: http://www.unoosa.org/oosa/en/ourwork/spacelaw/treaties/resolutions/res_16_1721.html (last visited on 25 Nov. 2018).

M. Chatzipanagiotis (✉)
University of Cyprus, Nicosia, Cyprus
e-mail: mchatz07@ucy.ac.cy

G. D. Kyriakopoulos, M. Manoli (eds.), *The Space Treaties at Crossroads*,
https://doi.org/10.1007/978-3-030-01479-7_6

6.1.1 Duty to Register Space Objects

A duty to register space objects was first established by the GA Resolution 1721B, which requests States launching space objects to furnish pertinent data to the UN for the registration of the launchings.

Details, however, were laid down in the RC, which constitutes *lex specialis* to the Resolution.[3] Art. II (1) RC obliges the 'launching State' to register in a special national registry space objects launched into Earth orbit. When there are more than one launching States, they need to arrange which of them shall register the space object [Art. II (2) RC]. At the same time, Art. IV RC obliges the launching State to furnish to the UN Secretary General certain information on the space object, including the name of the launching States, for registration in the special Register maintained.

The RC defines the launching State as (i) the State which launches or procures the launching of a space object and (ii) the State from whose territory or facility a space object is launched. Hence, there can be up to four launching States.

As a result, the State of registry must always be a launching State. International space law does not foresee the possibility to change the State of registry, yet Art. II (2) infers that such possibility cannot be denied among launching States.

6.1.2 Consequences of Registration

Registration of space objects aims at identifying the space object and connecting it with one or more States, to derive a series of consequences. Such consequences are foreseen in other international space law conventions, i.e. the Treaty on Principles Governing the Activities of States in the Exploration and Use of Outer Space, Including the Moon and Other Celestial Bodies (Outer Space Treaty [OST]),[4] the Convention on International Liability for Damage Caused by Space Objects (Liability Convention [LC]),[5] and the Convention on Registration of Objects Launched into Outer Space (Registration Convention [RC]),[6] which, combined with the RC, form the core of international space law (*corpus juris spatialis*).

---

[3] Thus, the Resolution has practical importance for States not having ratified the RC.

[4] *Treaty on Principles Governing the Activities of States in the Exploration and Use of Outer Space, Including the Moon and Other Celestial Bodies*, 27 January 1967, 610 UNTS 205, 18 UST 2410, TIAS No 6347, 6 ILM 386 (entered into force on 10 October 1967) [Outer Space Treaty].

[5] *Convention on International Liability for Damage Caused by Space Objects*, 29 March 1972, 961 UNTS 187, 24 UST 2389, 10 ILM 965 (1971) (entered into force 1 September 1972) [Liability Convention].

[6] *Convention on Registration of Objects Launched into Outer Space*, 6 June 1975, 28 UST 695, 1023 UNTS 15 (entered into force 15 September 1976) [Registration Convention].

6.1.2.1 Jurisdiction and Control

Art. VIII of the OST stipulates that the State of registry retains jurisdiction and control over the space object and over any personnel thereof, while in outer space, clarifying also that the ownership of the space object is not affected by its presence in outer space.

Nevertheless, according to Art. II (2) RC, the agreement among the launching States on the registration of the space object does not affect any agreements on the exercise of jurisdiction and control. Thus, Art. VIII OST refers to jurisdiction and control only vis-à-vis third States. Launching States are not necessarily bound by Art. VIII OST concerning their inner relationship.

The term 'jurisdiction and control' concerns the sovereign rights of the State of registry over the space object on the basis of non-territorial references, to take into account to prohibition of appropriation of outer space (Art. II OST).[7] 'Jurisdiction' refers to the right and authority to exercise legislative, executive and judicial power over the object and its personnel in outer space.[8] 'Control' has a technical meaning in the sense of the technical guidance and supervision of the crew activities.[9] The two notions are interconnected in that jurisdiction induces control and control is based on jurisdiction.[10]

The most important consequence of 'jurisdiction and control' is the applicability of the national law of the State of registry to occurrences regarding the space object and any persons thereon,[11] as well as the competence of its courts to hear the pertinent disputes. As a result, any judgements of foreign courts on disputes relating to the space object, in which foreign law has been applied, risk being unenforceable in the State of registry, owing to lack of international jurisdiction of these courts.

Nonetheless, Art. VIII OST does not establish exclusive jurisdiction of the State of registry over the space object. Other bases of jurisdiction, such as nationality of the persons involved, are not excluded. However, the jurisdiction of the State of registry has primacy over other forms,[12] in the sense that it is the reference point in the investigation of applicable law for occurrences on board a space object.[13]

---

[7] Bittlinger, Horst, *Hoheitsgewalt und Kontrolle im Weltraum*, Cologne et al. 1988, pp. 24–25.

[8] Vereshchetin, Vladlen, *Legal status of international space crews*, AASL 1978, 545 (547); Bordunov, V.D., *Rights of States as regards outer space objects*, 24 IISL-Proc. 1981, p. 89 (90).

[9] Vereshchetin, ibid., p. 548; Bordunov, ibid., p. 91.

[10] *Lafferranderie, Gabriel*, Jurisdiction and control of space objects and the case of an international intergovernmental organization (ESA), Zeitschrift für Luft- und Weltraumrecht 2005, p. 228 (231).

[11] *Schmidt-Tedd, Bernhard/Mick, Stephen* Art. VIII OST, para. 59, in: *Hobe/Schmidt-Tedd/Schrogl* (eds), Cologne Commentary on Space Law (CoCoSL), Cologne 2009.

[12] Gorove, Stephan, *Criminal jurisdiction in Outer Space*, Int'l Lawyer 1972, p. 313 (316–317); Jenks, C. Wilfred, *Space Law*, London 1965, p. 294.

[13] See details in *Chatzipanagiotis, Michael*, The legal status of space tourists in the framework of commercial suborbital flights, Cologne 2011, pp. 50–51.

The jurisdictional primacy of the State of registry can have practical consequences on disputes related to the financing of a space object, which might affect its operations, e.g. claims of the lessor against the operator/lessee in case of default of the latter. Furthermore, in case of manned operations, such as orbital hotels and commercial human spaceflight, the law of the State of registry will be the focal point regarding both criminal occurrences on board or in outer space and disputes on intellectual property rights (IPR) as to any inventions made.[14]

6.1.2.2 International Liability for Damage Caused by Space Objects

According to Art. VII OST, each State that launches or procures the launching of a space object and each State from whose territory or facility a space object is launched are internationally liable for any damage caused to another State.

The content of Art. VII OST has been specified in Arts II and III of the LC.[15] Art. II LC renders the launching State absolutely liable for any damage caused by its space object or to an aircraft in flight. Art. III LC clarifies that if the damage was caused elsewhere than on the surface of the Earth to another space object, then the launching State is liable only for fault. If there are more than one launching States, they shall be jointly and severally liable towards the State that suffered the damage (Art. IV LC).

The underlying reason of absolute liability according to the LC is that space activities are ultrahazardous; hence, the States conducting them should be liable irrespective of fault for damage caused on the surface of the Earth or to an aircraft in flight. Absolute liability for space activities is actually liability for injurious consequences arising out of acts not prohibited by international law, which has been studied by the ILC since 1978.[16]

6.1.2.3 Right to Have the Space Object Returned

The last sentence of Art. VIII OST provides that a space object shall be returned to the State of registry, if it is found beyond the territorial limits of the State of registry. According to Arts 4 and 5 (3) of the Rescue Agreement (RA),[17] if a space

[14] For a comprehensive analysis of these issues, see as to criminal law *Chatzipanagiotis, ibid.*, p. 39 et seq. and as to IPR *Bohlmann, Ulricke*, Kommerzielle Weltraumaktivitäten und die technischen gewerblichen Schutzrechte, Aachen 2002, p. 54 et seq.

[15] *Convention on the International Liability for Damage Caused by Space Objects*, adapted on 29 November 1971 by UN General Assembly Resolution 2777 (XXVI), in force since 1 September 1972.

[16] In 1996 the ILC decided to divide the topic into *Prevention of transboundary damage from hazardous activities* and *International liability in case of loss from transboundary harm arising out of hazardous activities* – see details at http://legal.un.org/ilc/summaries/9.shtml (last visited on 25 Nov. 2018).

[17] *Agreement on the Rescue of Astronauts, the Return of Astronauts and the Return of Objects Launched into Outer Space*, adopted on 19 December 1967 by UN GA Resolution 2345 (XXII), in force since 3 December 1968.

object and/or its personnel are found outside the territorial limits of the 'launching authority', they shall be returned to that State. Art. 6 RA defines the *launching authority* as the 'State responsible for launching'. The latter term corresponds to the notion of the launching State and is wider than the 'State of registry'.

6.1.2.4 International Responsibility for National Space Activities

Art. VI OST establishes the international responsibility of States for national space activities, irrespective of being conducted by governmental or non-governmental entities. The activities of the latter must be authorised and supervised by the appropriate State party. The scope of the provision is wide. It covers also States that are not 'launching States'.

The exact notions of 'national space activities' and 'appropriate State' are highly disputed. However, 'national space activities' is interpreted broadly and comprises all activities conducted from the territory of a State or by its nationals,[18] as well as activities conducted from ships, aircraft or space objects registered in its registries.[19] The definition of the 'appropriate State' should be interpreted as indicating the State that can exercise territorial and national jurisdiction over activities of private entities.[20]

International responsibility is closely connected with international liability under general international law, since liability is the obligation to compensate for damages.[21] If a State, because of faulty behaviour attributed to it, breaches an international obligation, then it is internationally liable to provide compensation for the damage produced, if the damage cannot be restored *in natura*.[22] Art. VI OST differs from general international law in that it assimilates activities of private persons with those of State organs.[23] Under Art. VI OST, there is direct State responsibility for the conduct of private persons.[24] As a result, a negligent breach of the duty to authorise and supervise national space activities may lead to international liability of the State.[25]

---

[18] *Cheng, Bin*, Article VI of the 1967 Space Treaty revisited: 'international responsibility', 'national activities' and 'the appropriate State', Journal of Space Law 1998, p. 7 (24–25); *Bittlinger, Horst*, Hoheitsgewalt und Kontrolle im Weltraum, Cologne 1988, p. 40; *von der Dunk, Frans*, Private enterprise and public interest in the european 'spacescape', Leiden 1998, p. 19; *Wassenbergh. Henri*, Principles of outer space law in hindsight, Dordrecht 1991, pp. 23, 27–28.

[19] *Cheng*, supra note 18, pp. 24–25; *von der Dunk*, supra note 18, p. 30.

[20] For a brief summary of the various views expressed on the interpretation of the term, see *Chatzipanagiotis*, supra note 13, pp. 133–135.

[21] *Gorove, Stephan*, Liability in space law: an overview, Annals of Air and Space Law 1983, 373 (374); *Kerrest, Armel/Smith, Lesly Jane* Art. VII OST, para. 4, in: *Hobe, Stephan/ von Rucketschell, Nikolai* (eds) Cologne Commentary on Space Law, Vol I, Cologne 2009.

[22] See *International Law Commission*, Draft articles on Responsibility of States for internationally wrongful acts (2001), adopted by UN GA Resolution 56/83 on 12 December 2001, which reflect mostly international customary law, especially Arts 2, 12, 31, 35 and 36.

[23] *Cheng*, supra note 18, p. 14; *Kerrest, Armel*, Remarks on the Responsibility and Liability for Damages caused by Private Activity in Outer Space, 40 IISL-Proc. 1997, p. 134 (139).

[24] *Cheng*, supra note 18, p. 15.

[25] See details in *Cheng*, supra note 18, 23, p. 8; *Gorove*, supra note 21, p.374.

It is also noteworthy that international responsibility and the arising liability under Art. VI OST is independent of the liability under Art. VII OST and the LC.

6.2 Commercialisation of Space Activities and Effects

Modern space activities include increasingly the commercial use of and access to outer space by private entities. Most prominent are the commercial uses of satellites for telecommunications, remote sensing and satellite navigation. At the same time, there is a great potential for commercial manned spaceflights serving recreational or scientific purposes, which include suborbital and orbital flights, flights to the International Space Station (ISS) and to future private space stations, while even flights to other celestial bodies, such as the Moon and Mars, are designed.

6.2.1 Modern Commercial Practices

The extensive commercialisation of space activities has brought about new commercial practices.

Transfer of ownership in orbit is not unusual. The operator of satellite in orbit transfers its ownership to another operator, which is subject to the law of another State.[26]

Turnkey satellite contracts aim at minimising the risk of the operator/buyer for a higher contractual price. An operator procures the construction and launching of a satellite but acquires ownership only when the satellite reaches the desired orbit. This way, the risk of delay in construction or in the launching of the satellite is borne by the satellite manufacturer.[27]

Moreover, satellites can be used as instruments to secure financing of the operators. In *leasing contracts*, the lessor/financier retains legal ownership of the asset, yet operational control is transferred to the lessee.[28] In the same vein, a *conditional sale* of a satellite does not allow ownership of the satellite to be transferred to the operator, until the operator has paid the full price. Besides, satellites can also be used as *security assets* for the operator's creditors, to facilitate its financing.

Hence, space objects are increasingly used as transactional objects, which results in legal ownership being separated from operational control.

---

[26] See examples of such cases in *Aoki, Setsuko*, in search of the current legal status of the registration of space objects, 53 IISL-Proc. 2010, p. 254 et seq.

[27] See details in *Ravillon, Laurence*, Les télécommunications par satellite, Bourgogne 1997, p. 139 et seq.

[28] In practice, however, leasing of a satellite as such is not very often. Leasing of transponder capacity is preferred instead.

6.2.2 Legal Ramifications

The above situations of separation between legal ownership and operational control entail a series of legal ramifications.

First of all, the State of registry, which has *de jure* the right to exercise international jurisdiction and technical control over the space object, is different than the State of operator, which has *de facto* control over the space object. The courts of the latter have no international jurisdiction over disputes related to the space object. Thus, any judgements they issue may have serious problems of recognition and enforcement in other States – despite any clauses on international jurisdiction and applicable law in the contract between the former and the new operator, because the validity of such clauses could be negated in the light of Art. VI OST.

In addition, the State of registry, as a 'launching State', bears absolute international liability according to Art. VII OST and the LC for any damage produced by the space object to third parties on the surface of the Earth or to an aircraft in flight – although it can exercise no factual control over the space object. On the contrary, the State of the operator is internationally responsible and could be held internationally liable only for negligence in authorising and supervising the operator according to Art. VI OST. Besides, the fact that more than one States can be a 'launching State' and that States are internationally responsible for national activities in outer space makes it difficult to safeguard themselves through their action, e.g. liability apportioning, indemnification agreements etc., inasmuch as jurisdiction and control is vested to the State of registry.[29] It should also be noted that as long as the State of the operator is not any 'launching State', no agreements are possible according to Art. II (2) RC, as it applies exclusively to 'launching States'.

Third, in case that the space object is found outside the territory of the launching State, it has to be returned to the State of registry, even if it is found in the territory of the State of operator. Yet, the State of registry would barely have any interest in the recovery of the space object.

6.2.3 Policy Considerations

Hence, complex legal situations arise, which can be addressed only through complicated agreements between the States concerned in combination with special provisions in national space legislation on authorisation and supervision of national space activities.[30] As a result, an amendment of the Space Treaties to streamline the actual with the legal situation appears necessary.

[29] *Cheng, Bin*, Space objects and their connecting factors, in: *Laferranderie G./Crowther Daphne (eds)*, Outlook on space law over the next 30 years, The Netherlands 1997, p. 203 (213).

[30] See, e.g. *Gerhard, Michael*, Transfer of operation and control with respect to space objects – Problems of responsibility and liability of States, Zeitschrift für Luft und Weltraumrecht 2002, p. 571 (577 et seq.); *Mick, Stephan*, Registrierungskonvention und Registrierungspraxis, Cologne 2007, p. 144 et seq.

Nevertheless, the question arises whether such need really exists. Especially regarding international liability, the Space Treaties are victim-oriented, and their aim is to identify at least a liable State that is connected somehow to the space object – which led the notion of the 'launching State'. Victims know exactly that at least the State of registry is liable for the damage they suffered. That State could then turn against the State of the operator, which is internationally responsible under Art. VI OST. Therefore, it could be argued that an amendment to the current system of international space law would jeopardise the victim-oriented character of the Space Treaties.

Such argument would not be solid, though. First, damaged persons can be protected adequately, as long as there is a clearly identifiable liable State. This need not be the 'launching State'. Second, complicated legal situations, especially the ones created by the discrepancy between legal presumptions and factual circumstances, tend to create international legal disputes and increase tension among States. Friendly relations between States and international cooperation, including indemnification of victims, are best served by clear provisions connected to the reality of the modern world. Third, the complicated solutions of the current legal regime increase transactional costs and hinder further technological development.

Therefore, amending the Space Treaties is the most viable solution.

6.3 The Aviation Model

The content of such amendments could be determined by examining the solutions adopted in aviation, which faced similar problems in the past, adjusted to the particularities of space operations.

6.3.1 Registration of Aircraft

Art. 17 of the Convention on International Civil Aviation (Chicago Convention [CC])[31] provides that aircraft bear the nationality of the State in which they are registered. There can be only one State of registry, yet change of registry is possible (Art. 18 CC). The registration or transfer of registration is made according to national law (Art. 19 CC).

In aviation law, registration follows nationality, which is determined by national law. However, the international aviation system imposes strict requirements on nationality. The principle of sovereignty of States in the airspace above their territory (Art. 1 CC) implies that national airspace is closed for foreign aircraft, unless there is a special agreement. Such agreements have been traditionally concluded bilaterally between States. To ensure that the bilateral agreements are

[31] *Convention on International Civil Aviation*, signed at Chicago on 7 December 1944, in force since 4 April 1947.

not circumvented, strict conditions on ownership and control of the aircraft are imposed, e.g. that the airlines involved have their registered seat in the contracting State and that the majority of their share capital is owned by nationals of the contracting States.

As a result, in aviation, nationality based on the genuine link principle determines registration. This principle, established by the International Court of Justice in the *Nottebohm* case,[32] stipulates that for the nationality of a person to have legal consequences under international law, there must be real factual ties between that person and the State of nationality.

Aircraft registration and nationality infer specific rights and duties to the State of registry, such as jurisdiction over the aircraft and occurrences on board, and international responsibility for safety oversight.

6.3.2 Art. 83bis CC

The growing commercialisation of international air travel led to the development of various commercial practices, in which legal ownership of the aircraft was separated from its operational control, e.g. leasing contracts, interchange agreements, charter flights etc.[33] The safety implications associated with the aircraft operator based outside the territorial jurisdiction of the State of registry led to the adoption of Art. 83bis CC through an additional Protocol to the CC.[34]

Art. 83bis allows the State of registry to agree with the State where the operator has its principal place of business (State of the operator) that certain duties and functions related to safety oversight will be transferred for a specific period to the latter State, which shall become internationally responsible therefor, while the State of registry will be relieved of the duties transferred.

6.3.2.1 Object and Purpose

The purpose of Art. 83bis is to facilitate safety oversight, while providing airlines with flexibility to agree commercial arrangements on the use of their aircraft.[35] It is a discretionary instrument at the disposal of ICAO Member States. The ratification of the pertinent Protocol does not entail automatic transfer of functions and duties – it requires a special agreement between the States concerned, which also identifies the exact aircraft it applies to.[36]

---

[32] Judgment of 6 April 1955 (*Liechtenstein* vs *Guatemala*), ICJ Reports 1955, p. 4.

[33] See on the content of such agreements *Manual on the Regulation of International Air Transport*, ICAO Doc 9626, 2nd ed., 2004, paras 4.1–8, 4.7–5 and 5.1–1.

[34] Protocol relating to an amendment to the Convention on International Civil Aviation signed at Montreal on 6 October 1980 [Article 83bis], in force since 20 June 1997.

[35] ICAO Cir. 295, para. 2.2.2.

[36] *Ibid.*, para. 2.2.3.

6.3.2.2 Content of the Agreement

The transfer of responsibility can cover only certain, albeit the most important, functions of the State of registry, i.e. respect of the rules of the air (Art. 12 CC), radio licencing (Art. 12 CC), certification of airworthiness (Art. 31 CC) and personnel licencing (Art. 32(a) CC).

The duties transferred as well as the aircraft concerned must be clearly and exactly identified, because the State of registry remains responsible for the rest.[37] Moreover, the agreement has to state the duration of the transfer, which usually corresponds to the duration of the commercial agreement, because no change of registration occurs.[38]

Consequently, Art. 83bis is a substitute for the change of registration. Although no change is undertaken, it amounts to a *de facto* partial change of registration, limited in scope and time. Other functions of the State of registry, such as participation in Search and Rescue operations and in accident investigation as well as the applicability of its law to occurrences on board aircraft, are not affected.

6.3.2.3 Formalities

The agreement can be invoked only as to States that have ratified the pertinent Protocol[39] and only after proper notification of these States.[40] Such notification can have the form of either registering the agreement with ICAO or directly informing the States possibly affected, e.g. the States to which the operator is commercially active.[41]

Consequently, the transfer of duties and functions to the State of the operator does not bind States non-parties to the additional Protocol and States that have not been properly notified, although they have ratified the Protocol.

6.3.3 Legal Consequences

Upon conclusion and proper notification of the agreement, the State of registry, for the period agreed, is relieved of its transferred duties and international responsibility and eventual liability rests with the State of the operator.

The implementation of such transfer may, however, prove challenging. The two States may have different provisions in force regarding the exact requirements

[37] *Ibid.*, paras 3.1, 3.3.

[38] *Ibid.*, para. 3.6.

[39] See Art. 83bis (2) CC combined with Art. 14 of the *1969 Vienna Convention on the Law of Treaties*, 23 May 1969, 1155 UNTS 331 (entered into force 27 January 1980) [Vienna Convention].

[40] Art. 83bis (1)(b) CC.

[41] ICAO Cir. 295 (*supra* note 35), para. 2.3.1.

aircraft operators have to fulfil. In addition, because the transfer concerns only certain functions and duties, the aircraft will be subject to two pieces of national legislation.

The situation is alleviated by the existence of uniform *Technical Standards and Recommended Practices*, which are practically binding to all ICAO Members under Art. 38 CC.[42]

6.4 Differences Between Aviation and Outer Space Operations and Regulation

Although the aviation model could be implemented in space law, significant amendments would be required, to take into account the differences in the legal and operational framework. Nonetheless, the operational conditions in outer space have begun to change, which means that regulation *de lege ferenda* should be far-sighted.

6.4.1 Legal Differences

Most importantly for the instances under consideration, aviation and space law differ as to the degree of legal sophistication. In aviation, detailed uniform technical rules on safety oversight, especially certification and licencing, are in force.[43] There are no such rules in international space law.

The detailed aviation rules enable the transfer of only certain duties and functions to the State of the operator according to Art. 83bis CC. However, jurisdiction and control under Art. VIII OST is a comprehensive notion, which does not permit so subtle distinctions.

In addition, registration in aviation is intertwined with nationality and the genuine link with the State of registry. In space law, registration is connected only to the notion of the launching State. Nationality of the persons on board a space object and ownership of the space object play no role under the RC.[44] Besides, the genuine link principle is incompatible with the current system of comprehensive liability imposed on the launching State(s).[45] Nonetheless, it has been submitted that where there is no registration, jurisdiction will probably follow ownership.[46]

---

[42] *Ibid.*, para 2.4.2.

[43] See, Annexes to the Convention on International Civil Aviation, e.g. Annex 1 Personnel Licensing, Annex 6 Operation of Aircraft, Annex 8 Airworthiness of Aircraft and Annex 19 Safety Management.

[44] *Cheng, Bin*, Space objects and their connecting factors, in: *Laferranderie G./Crowther Daphne (eds)*, Outlook on space law over the next 30 years, The Netherlands 1997, p. 203 (208).

[45] *Schmidt-Tedd/Mick* Art. VIII OST, para. 58, in: Cologne Commentary on Space Law, supra note 11; *Bittlinger, Horst*, supra note. 7, p. 76.

[46] *Bin Cheng*, Spacecraft, satellites and space objects in: *Bernhardt, Rudolf* (ed.) Encyclopedia of

As to liability, space law refers to the launching State, creating an irrebuttable (conclusive) presumption of the State liable towards third parties, to facilitate compensation.[47] The rationale is that space activities are ultrahazardous activities and States should bear the respective risk. In addition, the liable States should be easily identifiable; hence, the notion of the 'launching State' was created. Such strict liability is foreseen side-by-side with the compensation duty following the negligent breach of international responsibility to oversee national space activities. On the contrary, air travel is considered a routine activity, and the international duty to provide compensation is a corollary of international State responsibility for the negligent breach of oversight duties.

Besides, aircraft are subject of the territorial jurisdiction of the States in which they operate and the States overflown (Art. 1 CC). Space objects operate in outer space, which is not subject to the jurisdiction of any State (Art. I OST).

6.4.2 Operational Differences

Notwithstanding the differences in the legal regime, there are also operational differences between aircraft and space objects, mainly satellites.[48]

Uniformity: While the use of aircraft is relatively uniform (transport of passengers and cargo), satellites are designed for specific applications/missions (telecommunications in various bands, satellite navigation, Earth observation, space research, recreation, etc.).

Mobility and recoverability: Aircraft are very mobile and fly frequently across international borders and legal jurisdictions while being physically recoverable. Space objects operate in outer space and cannot be recovered physically.

Operational lifetime and control: Aircraft have an operational lifetime of a number of decades, their useful active life being extendable with replacement equipment and upgrades. They are often sold on, and many of them will have a number of different owners and operators over time. Operational control usually relates to safety aspects.

Satellites generally have a much shorter active lifetime. Transfer of ownership of space objects is not very usual. Operational control relates mainly to operation in orbit as well as to rights and obligations arising from registration and orbit spectrum. The fact that such rights are not readily assignable, combined with the special design of space objects for specific applications and the specialised expertise required to operate them, pose significant limits to the secondary market for a space object.

Public International Law, Vol. IV, Amsterdam et al. 2000, p. 557 (560).

[47] See Art. II of the Liability Convention.

[48] See Hughes, Nick, 'UNIDROIT Draft space protocol', online: http://www.hfw.com/UNIDROIT-Draft-Space-Assets-Protocol (last visited on 25 Nov. 2018).

Financing: Aircraft financing is primarily asset-based, whereas satellite financing is usually project-based relying on revenues flowing from use of the assets. Asset-based financing depends on there being a relatively long useful lifetime for the asset, on proper maintenance of the asset so that residual value is optimised and on being easily and swiftly repossessable should things go wrong. None of this is usually applicable to satellites.

Financial performance: The airline industry operates within a tightly regulated, customer-focused and low-profit margin environment, which limits its collective financial performance. There are numerous examples of operators becoming insolvent. This increases the inherent credit risk in, and the cost of, aircraft financing transactions.

By contrast, commercial satellite operators tend to enjoy much better financial results and a high level of profitability. This is a factor for satellite operators in being able to obtain finance based on the financial potential of the project, which effectively determines the value of the asset, rather than the residual value of the asset in an active market, which is the situation for commercial aircraft.

These operational differences indicate that in space law the equivalent of Art. 83bis CC would have very limited importance.[49]

6.4.3 *Gradual Change of Operational Concepts in Outer Space*

Until now the vast majority of commercial operations in outer space referred to the use of satellites, to provide terrestrial services. During the last years, however, new operational concepts have emerged.

Private human spaceflight has become a reality. Private entities offer orbital spaceflights to both private persons and governmental entities, which are currently being destined to the International Space Station (ISS).[50] Yet, private orbital and suborbital flights are envisaged for both recreational and scientific purposes in the near future.[51] There are also plans to conduct private spaceflights to the Moon[52] and Mars and even create Mars colonies.[53]

[49] This is also the reason why the space industry opposes the introduction of the Space Protocol to the Cape Town Convention, which aims at strengthening asset-based financing of space objects operators. See *Hughes*, supra fn. 48.

[50] For example, the *Dragon* vehicle of Space Exploration Technologies Inc. – see online: http://www.spacex.com/dragon (last visited on 25 Nov. 2018).

[51] For example, online: http://www.virgingalactic.com/human-spaceflight/; https://www.blueorigin.com/astronaut-experience (last visited on 25 Nov. 2018).

[52] For example, https://www.space.com/41854-spacex-unveils-1st-private-moon-flight-passenger.html (last visited on 25 Nov. 2018).

[53] See online: http://www.mars-one.com/ (last visited on 25 Nov. 2018), http://www.scientificamerican.com/article/martian-colony-spacex/ (last visited on 25 Nov. 2018).

Such flights are going to be combined with the entry into service of private space stations. These stations will be offered to anyone interested for various applications: States could use it for scientific research or for astronaut training; private entities and wealthy individuals could use it for scientific or leisure purposes (orbital laboratories and hotels). The architecture of the space stations is expected to follow the example of the ISS, e.g. different modules either standing alone or connected to each other. The same technology could be used to construct human and scientific habitats on celestial bodies.

In the long term, asteroid mining is contemplated.[54] There are studies indicating that asteroids are rich in mineral resources, which could have enormous value on Earth.[55] Supposing that the legal requirements for such operations will be established under international law, organising mining operations and transferring the extracted material on Earth will necessitate special mining equipment and transportation vehicles.

These concepts will possibly shift in part the economic and financial focus from services to assets. Space hotels, aerospace vehicles, mining robots and the like will have residual value, i.e. they will be valuable as such, not only because of the services they provide.

6.5 Proposal *de lege ferenda*

The above analysis shows that amending the Space Treaties would be necessary to address the legal challenges posed by the new technological developments. However, a complete overhaul of the Treaties would be neither desirable, because they still have many positive elements, e.g. a comprehensive system of protection of third parties, nor feasible, because States have proved extremely reluctant in amending the Space Treaties. Therefore, any suggested amendments will have greater chances of adoption, if they connect to the current rules in force.

6.5.1 Formal Registration and Liability

Registration of space objects and liability for damages caused by their operation should be disentangled from the notion of the 'launching State'. Instead, the respective rights and duties should be attached to the State of the operator: the State exercising effective operational control over the space object or the State authorising and supervising the non-governmental entity that exercises such control. The notion

[54] See, e.g. online: http://www.planetaryresources.com/ (last visited on 25 Nov. 2018).

[55] See, e.g. Keck Institute for space studies, 'Asteroid retrieval feasibility study' (2 April 2012) p. 12, online: http://www.kiss.caltech.edu/study/asteroid/asteroid_final_report.pdf (last visited on 25 Nov. 2018).

of effective control need not be similar to the strict *ownership and control* clauses in bilateral air traffic rights agreements, which require that both the majority of capital belongs to nationals of the contracting State and the majority of the board of directors are nationals of that State. The exact details of effective control could be left to national space legislation or to State practice, to ensure flexibility and thus greater acceptability of the new rules by the international community.

The State of the operator should be internationally liable for any damage caused during the operation of the space object. Liability of the 'launching States' under the LC could remain intact for the phases of the launch or initial operations of the space objects over which they exercise effective control.[56]

To facilitate the identification of the State of the operator and not compromise the protection of third parties, two rebuttable presumptions should be established. First, it should be presumed that the State of the operator is the State having registered the space object. Second, in case of space objects operated by non-governmental entities, it should be presumed that the State of the operator is the State of the place where the operator has its actual seat, i.e. the State where the operator's centre of interests is located and the most important management decisions are taken, hence, the place where effective control is exercised. Combining the two presumptions in practice would mean that third parties suffering damage from the operation of a space object could turn against the State of registry, which could be exonerated from liability if it proved that the operator's centre of interests lies in another State.

It is submitted that these provisions would achieve a better balance between the need to protect third parties and the interests of States involved in launching and operating a space object.

6.5.2 Change of Registration

Change of registration of the space object should be permitted, following a change of the State of the operator. This way, international responsibility and liability could be attached to the respective State of registry, which shall have both operational control and legal jurisdiction over the space object.

It could be argued that States should bear an international duty to proceed to re-registration of the space object following a change of operator. Nonetheless, international experience in aviation law has shown that such agreement need not be obligatory for the States concerned. It is possible that these States do not wish to alter the legal *status quo* of the space object for various reasons, e.g. to facilitate financing of operations of the space object, which could remain under a more creditor-friendly jurisdiction, or because operational control will be only short-termed.

---

[56] In this vein also *Lee, Ricky J.*, Transferring registration of space objects: The interpretative solution, IISL-Proc. 2000, p. 148 (153), who suggests that liability for space objects should be distinguished into three phases: launch, operation and retirement; in each phase the State that has operation and control over the space object should be liable.

If change of registration upon change of operational control is not obligatory, then there would be a risk that the State of operator does not accept an agreement on re-registration offered by the State of registry. However, depending on the circumstances, the refusal to sign such an agreement might be interpreted as a violation of Art. VI OST (failure to supervise national State activities) and entail international responsibility of the State of operator towards other States, including the State of registry.

In any case, de-registration of a space object should be allowed only if it is combined with re-registration by another State. This would minimise the possibility that a State proceeds unilaterally to de-registration of a space object, which would then appear strayed from a legal standpoint.

6.5.3 Registration of Bilateral Agreements

An alternative to the change of registration would be a bilateral agreement with third-party effect between the State of registry and the State of the operator, in the vein of Art. 83bis CC. Such agreement could regulate issues of liability and jurisdiction between the two States. Its purpose would be to facilitate the alignment of the legal situation with the factual circumstances created by a contract between the old and the new operator, e.g. leasing or conditional transfer of ownership.

As to liability, the State of registry could agree with the State of the operator that the latter assumes all international liability for damage caused by the space object to third parties. This could refute the presumed liability of the State of registry mentioned above. Another option would be that the State of the operator will be jointly and severally liable, as is currently the case with the liability of the launching States (Art. V LC).

Concerning jurisdiction, the State of the operator could be granted concurrent jurisdiction over the control stations of the space object that are subject to the territorial jurisdiction of the State of registry as well as its quasi-territorial jurisdiction, if they are outside the territorial jurisdiction of any State, e.g. the Antarctic, over the High Seas, or even in space or on other celestial bodies.

Furthermore, the State of the operator could obtain quasi-territorial jurisdiction over the relations between the operator and third parties – as opposed to the relations of the new operator/debtor with the old operator/creditor, which might be governed by the law of the State of registry, to secure the creditor's interests. Admittedly, the State of registry could retain jurisdiction through clauses on judicial jurisdiction (choice-of court) and the applicable law (choice-of-law) in the financing contract. Nevertheless, the courts of the State of operator might judge that the application of such clauses violates the *ordre public* of that State. This would result in applying the law of the State of registry instead of the law of the State of the operator, which could be potentially detrimental to the interests of the financier. Such risk would be alleviated, if jurisdiction is regulated by an international law instrument.

For example, imagine an orbital hotel operated by the operator A, who is supervised by State 1. The operator is financed by entity B, which has its actual seat in State 2 and had also arranged for the launching of the space hotel into Earth orbit. The hotel is registered by State 2. The relations of operator A with its customers and suppliers would be governed by the law of State 1. However, the relations between operator A and financier B would be governed by the law of the State 2. This way financier B would not have to worry about legal ramifications, if operator A defaulted, and financing of operator A would become easier. A provision on such agreements between the State of registry and the State of the operator could be inserted as Art. IIbis in the RC through an additional Protocol. The agreement should be able to be registered in the UN Registry and be invoked against third States that have ratified the Protocol.

6.6 Conclusion

In conclusion, the extensive commercialisation of space activities has resulted in sophisticated commercial practices that had not been contemplated by the drafters of the Space Treaties. This has resulted in legal rules that provide unrealistic and impractical solutions to the arising problems. The legal situation is expected to be further deteriorated by the future employment of new commercial practices, such as private space stations, private spaceflights and asteroid mining.

Therefore, the existing rules should be amended to reflect modern reality and future needs. Registration and liability should be attached to the State exercising effective control over the space object or over the non-governmental entity exercising such control. Change of registration should be possible upon agreement of the State having already registered the space object and the State of the operator. At the same time, States should be also given the option to conclude agreements on jurisdiction and control without changing the registration of the space objects, provided that such agreements are duly registered in the UN.

The amended international framework combined with pertinent provisions in national space legislation are expected to provide viable and appropriate solutions to legal issues of modern space activities and facilitate the development of the commercial space industry without compromising the protection offered to third parties.

Chapter 7
Is the Launching State the Only "Appropriate State" to Register a Space Object? Change of Registry in Case of Change of Ownership

Amalia Dimopoulou

7.1 Introduction

The United Nations treaties on the exploration and use of outer space ("space treaties")[1] date back to the beginning of the space era, when space actives were limited and generally carried out by only a few governmental actors. At the time international space law was called to reply to specific needs of the international community, such as to protect against the establishment of State sovereignty in outer space, to prohibit its use for non-peaceful purposes, to define the rights and obligations of States with regard to space objects. The regulation of private activities were addressed, but only as a subject of authorization and supervision of a certain State (the "appropriate State"), that should also bear international responsibility for them.

[1] *Treaty on Principles Governing the Activities of States in the Exploration and Use of Outer Space, Including the Moon and Other Celestial Bodies*, 27 January 1967, 610 UNTS 205, 18 UST 2410, TIAS No 6347, 6 ILM 386 (entered into force on 10 October 1967) [Outer Space Treaty]; *Agreement on the Rescue of Astronauts, the Return of Astronauts and the Return of Objects Launched Into Outer Space*, 22 April 1968, 672 UNTS 119, 19 UST 7570, TIAS No 6599, 7 ILM 151 (entered into force 3 December 1968) [Rescue Agreement]; *Convention on International Liability for Damage Caused by Space Objects*, 29 March 1972, 961 UNTS 187, 24 UST 2389, 10 ILM 965 (1971) (entered into force 1 September 1972) [Liability Convention]; *Convention on Registration of Objects Launched into Outer Space*, 6 June 1975, 28 UST 695, 1023 UNTS 15 (entered into force 15 September 1976) [Registration Convention]; *Agreement governing the Activities of States on the Moon and Other Celestial Bodies*, 5 December 1979, 1363 UNTS 3 (entered into force 11 July 1984) [Moon Agreement].

The views and opinions expressed are those of the author.

A. Dimopoulou (✉)
Centre National d' Etudes Spatiales, Paris, France
e-mail: amalia.dimopoulou@cnes.fr

G. D. Kyriakopoulos, M. Manoli (eds.), *The Space Treaties at Crossroads*,
https://doi.org/10.1007/978-3-030-01479-7_7

Since then, participation of non-governmental entities in space activities is ever increasing. Private entities often interact among them directly, without State intermediaries, and naturally they alter existing rights and obligations over space objects or establish new ones. By virtue of international rules of space law holding a State accountable for the performance of a space object, any change initiated by private entities might impact the relations of the respective States of their nationality.

The present paper shall focus to the shift of States' rights and obligations in the context of an in-orbit change of ownership, as seen under the light of the regime that surrounds the registration of space objects. A change of ownership essentially means a change of the entity carrying out the activity (operator) which equals to a shift of control over the space object. Whereas the State of registry is usually the State that supervises the activity, when a change of ownership occurs, a polarity between the State of registry and the State of Article VI of the Outer Space Treaty[2] is created.

The paper shall present the rules to be observed by the States involved to an in-orbit transaction, especially those of the Outer Space Treaty and the Registration Convention. Further, a mapping of the *de lege ferenda* obligations will be suggested.

7.2 States' Rights and Obligations in Relation to a Space Object Before and After an In-Orbit Transfer

The present part of this paper presents an overview of the rights and obligations of the State of registry[3] and the State that acquires ownership and control of the space object (State of transfer) under the light of the Registration Convention and the Outer Space Treaty. It also highlights inconsistencies arising following a change of the entity controlling the space object.

---

[2] Art. VI, Outer Space Treaty:

States Parties to the Treaty shall bear international responsibility for national activities in outer space, including the Moon and other celestial bodies, whether such activities are carried on by governmental agencies or by non-governmental entities, and for assuring that national activities are carried out in conformity with the provisions set forth in the present Treaty. The activities of non-governmental entities in outer space, including the Moon and other celestial bodies, shall require authorization and continuing supervision by the appropriate State Party to the Treaty. When activities are carried on in outer space, including the Moon and other celestial bodies, by an international organization, responsibility for compliance with this Treaty shall be borne both by the international organization and by the States Parties to the Treaty participating in such organization.

[3] According to Article I, c of the Registration Convention "[t]he term 'State of registry' means a launching State on whose registry a space object is carried in accordance with article II".

7.2.1 *Jurisdiction, Control, Authorization and Supervision Before In-Orbit Transfer*

Two fundamental rules govern States' rights and obligations in relation to the operation of a space object: first, every space activity must be carried out under State authorization and supervision (Article VI of the Outer Space Treaty) and second, a space object must be operated under the jurisdiction and control of the State in whose national registry they are carried (Article VIII of the Outer Space Treaty and Article II of the Registration Convention).[4]

(i) *Registration of space objects as a means to assert jurisdiction and control*

According to the Registration Convention, the launching State registers the space object in a national register[5] the establishment of which is informed to the Secretary-General of the United Nations (UN).[6] If there are two or more launching States, they shall commonly decide which among them is in the most appropriate position to exercise jurisdiction and control as the State of registry.[7] Registration must be effective throughout the lifetime of the space object.

Additionally, in order for the international community to be aware of the space objects launched into outer space and of the States linked to them, an international register of space objects launched into outer space is held by the UN Office of Outer Space Affairs[8] which is available online. Hence, full and open access is guaranteed to the information contained therein, in line with Article III of the Registration Convention.[9]

---

[4] The rules are also considered to have a customary nature; see Vladlen S Vereshchentin, and Gennady M. Danilenko, "Custom as a source of International Law of Outer Space" (1985) 13 (1) Journal of Space Law, 25.

[5] Registration Convention, Art. II: "When a space object is launched into earth orbit or beyond, the launching State shall register the space object by means of an entry in an appropriate registry which it shall maintain".

[6] Registration Convention, Art. II, para. 1: "When a space object is launched into Earth orbit or beyond, the launching State shall register the space object by means of an entry in an appropriate registry which it shall maintain. Each launching State shall inform the Secretary-General of the United Nations of the establishment of such a registry".

[7] Registration Convention, Art.I, para 2:

Where there are two or more launching States in respect of any such space object, they shall jointly determine which one of them shall register the object in accordance with paragraph 1 of this article, bearing in mind the provisions of article VIII of the Treaty on Principles Governing the Activities of States in the Exploration and Use of Outer Space, including the Moon and Other Celestial Bodies, and without prejudice to appropriate agreements concluded or to be concluded among the launching States on jurisdiction and control over the space object and over any personnel thereof.

[8] See "Online Index of Objects Launched into Outer Space", online at: UNOOSA http://www.unoosa.org/oosa/osoindex/search-ng.jspx?lf_id (last accessed 15 March 2018).

[9] Registration Convention, Art. III: "1. The Secretary-General of the United Nations shall maintain a Register in which the information furnished in accordance with article IV shall be recorded.2. There shall be full and open access to the information in this Register".

In accordance with Article VI of the Registration Convention, the State of registry notifies as soon as practicable the Secretary-General of the UN of the launch and provides the following information[10]: (a) the name of launching State(s); (b) an appropriate designator of the space object or its registration number; (c) the date and territory or location of launch; (d) certain basic orbital parameters including nodal period, inclination, apogee and perigee; and (e) the general function of the space object. The information can be amended throughout the lifetime of the space object.[11]

Pursuant to Article VIII of the Outer Space Treaty "a State Party to the Treaty on whose registry a space object launched in outer space is carried shall retain jurisdiction and control over such object and over any personnel thereof".[12]

Jurisdiction in public international law derives from State sovereignty and is the capacity of a State to exercise certain powers over natural and legal persons as well as over property. In other words it refers to the power of a State to regulate or impact people or circumstances.[13] In the law of outer space, jurisdiction under the scope of Article VIII of the Outer Space Treaty is understood as the authority of the State to exercise certain powers over the space object. On the other hand, "control" is perceived as the "strictly technical element of jurisdiction",[14] meaning operational control. Needless to say that the (governmental or not governmental) entity asserting control of a space object practically has immediate and direct influence over it, as the sole entity in a position to programme, manoeuvre and operate it in any way.

Whereas the act of registration at national level creates a legal link between the State and an object found in a place escaping territorial jurisdiction,[15] the obligation to inform the Secretary-General of the UN calls for transparency of this link.

---

[10] See Registration Convention, Art. VI.

[11] Registration Convention, Art. IV, para.2: "Each State of registry may, from time to time, provide the Secretary-General of the United Nations with additional information concerning a space object carried on its registry".

[12] Outer Space Treaty, Art. VIII:

A State Party to the Treaty on whose registry an object launched into outer space is carried shall retain jurisdiction and control over such object, and over any personnel thereof, while in outer space or on a celestial body. Ownership of objects launched into outer space, including objects landed or constructed on a celestial body, and of their component parts, is not affected by their presence in outer space or on a celestial body or by their return to the Earth. Such objects or component parts found beyond the limits of the State Party to the Treaty on whose registry they are carried shall be returned to that State Party, which shall, upon request, furnish identifying data prior to their return.

[13] Malcolm N Shaw, *International Law* (Oxford: OUP, 2006), 645.

[14] Niki Aloupi, "Le rattachement des engins à l'Etat en droit international public (navires, aéronefs, objets spatiaux) » (2011) Thèse de doctorat, Université Panthéon-Assas, 264.

[15] See Bernhard Schmidt-Tedd/ Michael Gerhard, "Registration of space objects: Which are the advantages for States resulting from registration?" in Marietta Benko/ Kai-Uwe Schrogi, eds, *Space Law: Current problems and Perspectives for future regulation* (The Netherlands: Eleven International Publishing, 2005), 124–129; Angelos Yokaris, *International Air and Space Law* (title translated from Greek) (Athens: Sakkoulas Publishing, 1996), 256–257.

Registration in the UN registry indicates the State supposed to have the closest link to the space object, in other words, the State that shall be liable to pay any damage caused by this space object[16] by means of the Liability Convention.

(ii) *Authorization and supervision by the "appropriate State"*

Article VI of the Outer Space Treaty[17] establishes a fundamental rule of international space law invoking the international responsibility[18] of a State for the activities of its private entities in outer space: "States shall bear international responsibility for national activities in outer space, whether such activities are carried on by governmental agencies or by non-governmental entities".[19] For this reason, the activities of non-governmental entities in outer space "shall require authorization and continuing supervision by the appropriate State".[20]

The term "appropriate State" has concerned many scholars since it was introduced in the Outer Space Treaty, where it substituted the term "concerned State" of the United Nations Declaration of 1965.[21] In general terms it has been argued that the State of nationality or the State responsible for the activity or the launching State or any State involved to the launch[22] could qualify as the "appropriate State". However, since Article VI sets no other criterion than the mere act of authorization and supervision of a space activity, it suffices to spot the State best qualified best to do so.[23]

Furthermore, it is up to each State to define the terms and conditions under which it authorizes a space activity and thus becomes the "appropriate State",[24] as both authorization and supervision of space activities constitute an integral part of State sovereignty. Authorization usually takes the form of operators' licencing.[25] Licence granting practically consists on a governmental validation that the activity is

---

[16] Philippe Achilléas, "Le droit international de l'espace » in Philippe Achilléas (Dir.), *Droit international de l'espace: Télécommunication- Navigation-Défense- Exploration* (Paris: Larcier, 2009), 30.

[17] See Outer Space Treaty, Art. VI.

[18] The French term is « responsabilité de contrôle », responsibility to control; see Achilléas, p. 27.

[19] See Outer Space Treaty, Art. VI.

[20] Ibid.

[21] *Declaration of Legal Principles Governing the Activities of States in the Exploration and Use of Outer Space*, UN GA Res A/RES/18/1962, 13 December 1963, para. 5.

[22] See, generally, Julien Hermida, *Legal Basis for a National Space Legislation* (New York: Kluwer, 2004).

[23] Ricky J. Lee, "Liability Arising from Article VI of the Outer Space Treaty" in Proceedings of the 48th Colloquium on the Law of Outer Space, 17–21 October 2005 Fukuoka, AIAA, 2006, 216.

[24] When States undertake legislation to this end they invoke Article VI as the legal basis. National space laws aim to flow down to national entities the obligations set out in the space treaties; Alain Costes, *L'évolution du droit de l'espace en France* (France: Ministère délégué à la recherche et aux nouvelles technologies, 2002), 23.

[25] For example, in France, the obligation of an operator to have a licence depends on whether the State exercises personal (nationality) or territorial (establishment) jurisdiction; see French Space Operations Act 2008–518, 3 June 2008, Art. 2.

compliant with national standards and, as soon as it takes place, the State accepts to continually supervise the activity.[26] Having said that, contrary to what has been argued by certain commentators[27] and in line with the wording of the Outer Space Treaty,[28] there can only be one appropriate State[29] for each space activity.

Due to the similar content of the rights of the State of registry (jurisdiction and control over the object) and the appropriate State (jurisdiction over the private entity to authorize and continually supervise its activities), it is usual, but not obligatory, that these two qualifications incorporate to the same State and that the registry State is also the "appropriate State". However, even if this might be the case before in-orbit transfer, it is certainly not after it. In the next section we shall see why.

7.2.2 *Mapping of Rights and Obligations over the Space Object After the Change of Control*

Once a space object operating in-orbit is transferred from a private entity to another, if the entity acquiring it is not authorized by or established in the State of registry, a second State, the State of nationality or establishment of the operator, may interfere with the exercise of rights and obligations by the State of registry. In order to better understand the context of in-orbit transaction at State level, the following example shall be introduced:

Supposing that State A has launched and registered according to the Registration Convention Satellite S, owned and operated by Operator A, a national non-governmental entity of State A carrying out space activities under a general licence by the same State. A specific licence for the operation of S has also been issued by A. A few years later, Operator A proceeds to transfer of the ownership of S to Operator B, a national of State B also licenced under the national law of State B.

After the transaction, control and jurisdiction, presumed to be exercised by the same State, are dissociated[30]: operational control of the space object is transferred to a new entity, national of the State of transfer (Operator B), whereas jurisdiction remains with the State of registry, State A. The latter finds itself unable to comply with the obligation to exercise exclusive[31] jurisdiction and effective control during

[26] Frans Von der Dunk, "The Origins of Authorisation: Article VI of the Outer Space Treaty and International Space Law" in Frans Von der Dunk, ed, *National Space Legislation in Europe* (Germany, Brill: 2011) 17–18.

[27] Bin Cheng, "Article VI of the Space Treaty Revisited: "International Responsibility", "National Activities" and "the Appropriate State", Journal of Space Law (1998) 26(1), 28–29; Hermida, *supra* note 23, 43–44.

[28] Francis Lyall and PaulB Larsen, *Space Law – A Treatise* (England: Ashgate, 2009), 469.

[29] Von der Dunk, *supra* note 27, 17–18.

[30] Aloupi, *supra* note 15, 265–267.

[31] Aloupi, *supra* note 15, 384.

the whole lifetime of the satellite in-orbit.[32] Failure to exercise exclusive jurisdiction and efficient control consists on a breach of Article VIII of the Outer Space Treaty, resulting to international responsibility of State A under the rules of public international law.[33]

On the other hand, even if State B has operational control, the exclusive character of the registry State's jurisdiction prohibits the exercise of similar powers[34] over the space object. Any attempt of State B to interfere with the operation of the space object through Operator B might be interpreted as an intervention to the internal affairs of State A.

In terms of Article VI of the Outer Space Treaty, supervision is transferred from State A to State B. The State of transfer exercises personal jurisdiction over the new operator and has the obligation under Article VI of the Outer Space Treaty to continuously supervise its activities. The moment the space object enters the new operator's control, its operation enters the sphere of the State of its nationality, bound to ensure continuous supervision of its space activities, to which operation of S has been added.

After the change of control, the State of transfer not only has effective control of the satellite but also the obligation to authorize and supervise the activities of its operator. Hence, between the State of registry and the State of transfer, it is the second in best position to assert jurisdiction both over the operator and the satellite. It follows that, after the shift of effective control from an operator to another and the subsequent change of supervision from a State to another, the State of transfer becomes the "appropriate State" under Article VI of the Outer Space Treaty.

The new status quo might create further confusion throughout the intentional community because the State asserting real control is not the one appearing as State of registry. Further, in case of damage caused by the space object, the State of registry shall be directly identified as the State having an effective link to it, and compensation claims will be directly addressed to it. The State of registry will then compensate and turn to State B to claim its own damages, as it would have done with other launching States. State B however could refuse to compensate claiming that a non-launching State cannot be held liable.[35]

A legally safe solution would be for jurisdiction to follow operational control and for the international community to have enough information to comprehend the new plexus of rights and obligations exercised over the space object. The current legal regime provides with some tools to deal with the situation. Should those tools be found not entirely satisfactory, a revision should be considered.

---

[32] See Outer Space Treaty, Art. VIII.

[33] ILC Articles on the Responsibility of States for Internationally Wrongful Acts, Official Records of the General Assembly, 56th Session, Supp. No. 10, U.N. Doc. A/56/10; Schmidt-Tedd, *supra* note 16, 137.

[34] Shaw, *supra* note 14, 651.

[35] See Liability Convention, Arts. II and III.

7.3 Considerations *de lege lata* and *de lege ferenda*

As a remedy to dissociation of jurisdiction and control, jurisdiction should be shifted from the State of registry and exercised exclusively by the State of transfer. Once there is an agreement on the specific rights and obligations of each State, adequate and timely information should be provided to the UN reporting the changes worth to be known to the international community.

Section A below provides an overview of the relevant rules and their application to an in-orbit transfer, whereas Section B proposes a way forward in reviewing the Registration Convention in order to cover cases where a State third to the initial launch acquires rights over a space object.

7.3.1 The Rules to Bring Balance

(i) *Transfer of registry, jurisdiction and liability*

As explained in the previous section, it is possible for a State of registry to be deprived of operational control in favour of a third State essentially involved in the operation of the space object. Taking under consideration Article II of the Registration Convention providing that registration must be effective throughout lifetime of the space object, the quickest way to restore the balance of rights and obligations over the space object would be to change the State of registry. This is however not possible in all cases.

The decisive element for a State to qualify as a registry State[36] is participation to the initial launch, either as the State that launched or procured the launch or the State that provided its facilities or territory for the launch.[37] In the example of Section B, supposing that State B is a launching State, change of control can be followed by a change of registry technically without complications: State A shall eliminate the space object from its registry and State B shall register it,[38] as soon as all launching States agree to it (Article II, para. 2 of the Registration Convention). If however State B is not a launching State, transfer of registry is not allowed.

Even when transfer of registry is not possible, there are still tools to reunite jurisdiction and control under one State. Article II, para. 2 of the Registration Convention provides that the decision concerning the determination of the State of registry is made by the launching States by taking into account the elements of jurisdiction and control and "without prejudice of appropriate agreements concluded or to be concluded among the launching States on jurisdiction and control over the space object

[36] Registration Convention, Art I, para. c "The term 'State of registry' means a launching State on whose registry a space object is carried in accordance with article II".

[37] See Liability Convention, Art. 1, para a.

[38] Stephan Hobe, et al., eds, *Cologne Commentary on Space Law, Vol II*(Germany: Carl Heymanns Verlag, 2013), 256; Hermida, *supra* note 23, 65.

and over any personnel thereof".[39] This provision has been interpreted as permitting ad hoc interstate agreements for the transfer of jurisdiction over a space object. In accordance to Article II, para. 2, the State of registry as holder of jurisdiction, the other launching States and the State of transfer should agree in written between them to transfer jurisdiction.[40]

A similar approach has been adopted regarding liability. By virtue of Article V of the Liability Convention[41] under condition that there is an agreement of all launching States, the obligation to compensate can be transferred to a third State, even if it is not involved in the launching event, if this State accepts to undertake such an obligation.[42]

Once decisions regarding transfer of jurisdiction and liability are taken, the State of registry should inform the international community about any changes of the initial entry to the UN registry.

(ii) *The transparency requirement*

Article VI of the Registration Convention introduces international registration as an action with a certain dynamic and not as a one-time event, stipulating that the State of registry provides, from time to time, additional information to the Secretary-General of the UN. If States mean to respect this provision forming the *raison d'être* of registration in the UN registry, the transfer of control and/or jurisdiction over a space object is by default an event to be reported to the United Nations. Provision of additional information is also addressed in the United Nations General Assembly Resolution 62/101.[43] Information to be reported includes the name of the new operator, any change on supervision, any changes in orbital position, change of function and other relevant characteristics.

Resolution 62/101 and General Assembly Resolution 68/74[44] were both voted with the purpose to provide with guidelines to national legislation in the context of the evolution of private space activities. They provide soft law obligations[45]

---

[39] Registration Convention, Art. II, para. 2:

Where there are two or more launching States in respect of any such space object, they shall jointly determine which of them shall register the object [...] bearing in mind the provisions of Article VIII of the [Outer Space Treaty] and without prejudice of to appropriate agreements concluded or to be concluded among the launching States on jurisdiction and control over the space object and over any personnel thereof.

[40] Kay-Uwe Schrogl and Julien Hermida, "Change of Ownership, Change of Registry? Which Objects to Register, What Data to Furnish, When and Until When?", in Proceedings of the 46th Colloquium on the Law of Outer Space, Bremen 2003, AIAA, 2006; Hermida, *supra* note 23, 65.

[41] Liability Convention, Art. V: "[t]he participants in a joint launching may conclude agreements regarding the apportioning among themselves of the financial obligation in respect of which they are jointly and severally liable".

[42] Aloupi, *supra* note 15, 333.

[43] *Recommendations on enhancing the practice of States and international intergovernmental organizations in registering space objects,* UN GA Res. A /RES/62/101(10 January 2008).

[44] *Recommendations on national legislation relevant to the peaceful exploration and use of outer space*, UN GA Res A /RES/68/74 (16 December 2013).

[45] Shaw, *supra* note 14, 285–386.

("recommendations on national legislation") aiming to the establishment of national laws to harmonize and facilitate the exercise of State rights over space objects operated by non-State actors. Resolution 62/101 highlights the importance of collaboration of the two States when it comes to informing the UN.[46] Further, Resolution 68/74 suggests that by means of national legislation, the national authority holding the registry should request by the operator/owner or by "the State responsible for national activities in outer space" (this meaning the appropriate State) to furnish information in order to facilitate submission of any necessary elements to the Secretary-General of the UN.[47]

The two resolutions aim to a "most complete registration" of space objects in the context of the increase of space activities carried out by non-governmental entities. It is clear that the obligation to inform remains with the State of registry which has to be assisted by the State of transfer. Fairly so, after the change of control, the State of registry might not have access to all the necessary information to be reported to the UN and the State of transfer has no obligation to inform whatsoever. Only in case that "there is no State of registry, the appropriate State according to article VI of the Outer Space Treaty could furnish the above information to the Secretary-General".[48]

Even though the rules in force provide tools allowing to restore at least a degree of legal certainty as per the rights exercised over a space object, plenty of room is left to State discretion as to the approach to be followed. The tools described in this section are of limited effect as the proposed solutions can be applied on a case-by-case basis, depending on the choices of the actors involved. The evolution of commercial space activities invokes the need to come with regulatory and generic solutions addressing the issue in a more consistent way.

7.3.2 Transfer of Registry: Revisiting the Registration Convention and the Outer Space Treaty

Ex post involvement of a non-launching State to the operation of a space object is a novelty that ought to be addressed by the *corpus iuris spatialis*. Regardless of any ad hoc State agreements, international law needs a concrete method of attribution of rights when an already launched space object becomes subject of transaction between nationals of different States.

Legal certainty implies that registration should not only effective but also as accurate and complete as possible. Neither a State of registry with limited rights nor an appropriate State with no jurisdiction can be.

[46] Resolution 62/101, *supra* note 44, para. 4.

[47] Resolution 68/74, *supra* note 45, para. 6.

[48] Resolution 62/101, *supra* note 44, para. 4(b).

As argued in the previous sections, the State of transfer exercises not only control but also authorization and supervision and is, from a legal and an operational point of view, in a position to affect the space object more than any of the launching States, the State of registry included. Since the State of transfer substitutes the State of registry in so many ways, it only seems natural that it carries the space object in its registry, regardless of its qualification as a launching State.

A teleological interpretation of the Registration Convention also leaves room for allowing the above, provided that the State of transfer compiles with the requirement of jurisdiction and control. Registration of space objects is put in place so as to indicate which State has the most effective link with a certain space object. When the link to the initial State of registry is lost, as it is the case after an in-orbit transaction, a new effective link should be established between the space object and the most appropriate State to exercise jurisdiction and control, the State of transfer. *De lege ferenda*, the State of transfer should be entitled to register the space object as soon as the shift of control takes place, whether it is a launching State or not.

On the moment the treaties where drafted, the act of launching a satellite in outer space was seen as the sole significant event justifying the creation of a legal link between a State and a space object.[49] Increasing privatization of activities in outer space and subsequent transfers of the ownership of a space object prove that the launching event is not the only way to establish a link between a State and an object. On the contrary, through authorization and supervision of the operator, any State can potentially acquire a link to the space object. It is important that the Registration Convention evolves so as to accommodate this reality.

7.4 Conclusion

To conclude, in-orbit transaction of a space object results not only to its change of control but also to change of supervision over the activity. The state of transfer, as the new State that authorizes and supervises the operation of the space object, becomes the appropriate State of Article VI of the Outer Space Treaty. The State of registry on the other hand, deprived from operational control, is no longer in a position to comply with its obligations, and, as soon the UN Secretary-General is informed, it should be released from them, provided that another State is willing to register. It has been shown that the most appropriate State to register the space object in this case is the State of transfer. The current legal regime however, limits the States eligible to register to launching States.

In order to adapt to the increasing commercialization of outer space activities, the *corpus iuris spatialis* should evolve towards a direction where States third to the initial launch can, under certain circumstances, acquire rights and obligations that are now reserved only to the launching States. For instance, the State of transfer

[49] Yokaris, *supra* note 16, 266.

should, *de lege ferenda*, be allowed to register a space object found under its authorization and supervision, as long as it asserts jurisdiction and control over it and the initial registry State agrees to the change of registry.

Chapter 8
From Sea to Outer Space and Back: Political, Economic, and Environmental Considerations for Ocean-Based Space Launching Activities

Gerasimos Rodotheatos

8.1 Introduction

The connection between space activities and the marine domain exists since the early days of the Space Era. Auxiliary tasks such as position tracking, telecommunications, and space object recovery conducted by vessels and installations were already part of the routine since the 1960s.[1] Eventually, various experimental and operational rocket launchings were being carried out by vessels, submarines, as well as platforms, though mainly suborbital. During the last decades, a new trend has emerged that entails not only launching but also landing of space objects from and at the seas (hereafter "Ocean-Based Space Activities" – OBSAs). Such a kind of activity involves various advantages in financial, safety, and operational terms, but since it has not been widely practiced or studied, there are certain potential threats and knowledge gaps that ought to be examined further on. The scope of this paper is to highlight the prospects and dangers for Launching States on the one hand and on the other to emphasize on the priorities and rights of Coastal States and other ocean users.

8.2 How Does it Work?

Launching and landing are complex and multileveled activities that demand precision and fine-tuning. Moving out to the oceans adds up another dimension, since different systems' configurations ought to be deployed, safety standards

[1] See Chapter 6, "NASA Historical Data Book: Volume III – Programs and Projects 1969–1978" SP-4012, online: NASA http://history.nasa.gov/SP-4012/vol3/ch6.htm.

G. Rodotheatos (✉)
European Centre for Environmental Research and Training,
Panteion University, Athens, Greece
e-mail: yrodo@panteion.gr

G. D. Kyriakopoulos, M. Manoli (eds.), *The Space Treaties at Crossroads*,
https://doi.org/10.1007/978-3-030-01479-7_8

change, and unusual conditions have to be tackled. A brief study shows that material, software and hardware, and payloads launched or landed in the oceans are identical to those used for continental activities, but not without major or minor alterations.

A typical OBSA configuration mainly contains a command and control base, a tracking–telecommunications segment, a launch pad, and one or more hangars. For the present analysis, of critical importance is the type of platform that carries the launch pad, as this defines its position and relation to the marine domain. Till today, there are three categories of platforms that have hosted launch pads in OBSAs: (a) fixed installations, (b) floating installations, and (c) submarines.

Fixed installations are constructions resting upon the seafloor by means of piles or tubes driven into the bottom, or concrete structures.[2] *Floating installations* are identical to fixed, in terms of shape and material, but instead of resting, those structures float on the water column, and their position is secured either mechanically (Dynamic Positioning) or through anchoring,[3] while *submarines* are all vessels designed to operate below the surface of the sea.[4]

From a law of the sea point of view, fixed and floating installations hold the status of Artificial Islands and Structures[5] (hereafter AIS), while submarines are considered to be vessels. The Law of the Sea Convention (hereafter "LOSC")[6] contains nothing less than 12 different versions that fall under the AIS category, depending on use and position. According to our analysis, installations used in OBSAs are mainly regulated by Article 60, Para. 1 (referred as "c. *installations and structures* which may interfere with the exercise of the rights of the coastal State in the zone"); 147, Para. 3 (referred as "*Other activities* in the marine environment shall be conducted with reasonable regard for activities in the Area"); and 87, Para. 1 (referred as "d. freedom to construct *artificial islands and other installations* permitted under international law"). Submarines on the other hand are not thoroughly regulated by LOSC, which contains just a few relevant articles, since most of their activities are covered by Law of Armed Conflict rules[7].

This paper will use as study sample five OBSAs, in which platforms covering the full spectrum of categories are being engaged:

1. *Luigi Broglio Space Center* (formerly known as "San Marco Equatorial Mobile[8] Range"). An Italian-led initiative, first of this kind, carried out with the techno-

[2] Al. Soons, "Artificial Islands and Installations in International Law", Occasional Paper no 22, Law of the Sea Institute, University of Rhode Island, 1974, 3.

[3] S. Chakrabarti, *Handbook of Offshore Engineering*, vol 1, (Amsterdam: Elsevier, 2005), 3.

[4] *Treaty for the Limitation of Naval Armament*, London, 1936, Art. 1, para. B.4.

[5] G Walker, ed., *Definitions for the Law of the Sea Terms Not Defined by the 1982 Convention*, M., (Leiden: Nijhoff, 2012) 104–107.

[6] *United Nations Convention on the Law of the Sea*, (10 December 1982), entered into force 16 November 1994.

[7] J. Jacobson, "The Law of Submarine Warfare Today" in H. Robertson, ed., *International Law Studies – The Law of Naval Operations*, vol. 64 (1991), 205–240.

[8] "The project San Marco will be performed (…) from a mobile base consisting of two floating platforms with movable legs;" see H. Nesbitt, *History of the Italian San Marco Equatorial Mobile Range*, NASA CR-111987, 1971, 5. Even though both installations (San Marco and San Rita) were capable of relocating, no relocation was ever reported.

logical support of US agencies. It includes two fixed installations, San Marco and Santa Rita, resting offshore the Kenyan coast, both hosting launch pads.[9] Even though it was considered to be a successful endeavor,[10] launching activities lasted for only about 25 years. The Space Center is still in action, but there are no operations planned for the future.[11]

2. *Sea Launch.* A previously Swiss-based multinational and now turned Russian enterprise,[12] with an operational base in Long Beach, California. Its launch platform "Odyssey" is a self-propelled floating installation, usually positioned in the Pacific Ocean. Active since 1999 and with a high success rate, it is temporarily out of service and is expected to resume operations in 2019.[13]
3. *SpaceX.* A US private firm that apart from the development of launch vehicles and launching operations has also created a reusable rocket boosters' landing system. Up to date the company has deployed three Autonomous Spaceport Drone Ships (named "Just Read the Instructions" and "Of Course I Still Love You"[14]), exclusively used for ocean-based landings. The first two floating

[9] The initial planning was for Santa Rita to serve as an auxiliary platform. Instead, it also served as a launch pad for a small number of Nike-Apache rockets in 1964.

[10] Carroll considers the San Marco Program as a first proof of OBSAs' viability; see J Carroll, "From the Seas to the Stars – A Case for Developing Offshore Spaceports on States' Submerged Lands," *William & Mary Environmental Law and Policy Review* 39(3), 766. Italy was also the first space-faring nation, after USSR and the USA, that launched a satellite (San Marco B) from its own facility (the San Marco Equatorial Mobile Range) in April 1967; see G Palumbo, "From (under)ground to space. The birth of Space Science in Italy and the growth of astrophysics in its National Institutions" in *Nuclear Physics B (Proceedings Supplements)*, vol. 212–213, (2011), 42.

[11] In 2004 the Italian Space Agency and Russia examined the possibility of using the L. Broglio Centre, by the latter. The preliminary findings were characterized as "positive" by ASI; see ASI, *Piano AeroSpaziale Nationale 2006–2008*, 69. In 2016 Kenya and Italy decided the renewal of their cooperation regarding L. Broglio Centre, but none of the priorities set included launching of space objects; see "Via libera del Governo all'accordo col. Kenya," online: ASI http://www.asi.it/it/news/il-governo-ratifica-laccordo-col-kenya (last accessed 15/01/2018).

[12] The stakeholders of the initial scheme (Sea Launch Co. LLC, 1995–2010, registered in Cayman Islands) comprised of a US company (Boeing Commercial Space, 40%), a Russian company (Energia, 25%), a Norwegian company (Aker Solutions, 20%), and a Ukrainian one (SDO Yuzhnoye/ PO Yuzhmash, 15%). After a restructuring, a new synthesis emerged in October 2010: Energia took hold of 95% and Boeing C.S. and Aker 2.5% each, while SDO Yuzhnoye/ PO Yuzhmash withdrew (Source: http://www.sea-launch.com). In 2017 the Russian company S7 Group took over Sea Launch assets, while it signed a cooperation agreement with Energia on engineering, launch support, and system integration services. See "S7 Group announcing an Agreement to Purchase Sea Launch"; online: Sea Launch http://www.sea-launch.com/news/11421; "S7 Group has become an aerospace holding company"; online: S7 https://www.s7.ru/home/about/news/s7-group-has-become-an-aerospace-holding-company (last accessed 15/01/2018).

[13] According to press sources, this new scheme aims at conducting 10–12 launches in the first 5 years, starting from 2019; see "Group intends to resume launches from Sea Launch floating spaceport in 2019"; online, Ruaviation https://www.ruaviation.com/news/2017/12/13/10364/print/ (last accessed 15/01/2018).

[14] All platforms where initially used as barges in various maritime projects and were refitted by SpaceX. "Just Read the Instructions" ("JRTI") was built in 2015 and was followed by "Of Course I Still Love You" ("OCISLY") a few months later. "JRTI" was used only for a few months

platforms have operated in the Atlantic Ocean, while the third one[15] has been deployed in the Pacific Ocean. After a series of failures, the first successful landing took place in April 2016 in the Atlantic Ocean, and since then the results have been quite encouraging.

4. *Copenhagen Suborbitals.* A Danish nonprofit organization aiming at developing and launching low-budget vehicles and devices and that has chosen to perform only marine-based launches.[16] To this scope, it has constructed the Mobile Launch Platform "Sputnik," a floating installation being towed to the appointed launch sites (located in the Baltic Sea or the Danish Straits) when needed. MLP Sputnik is operational since 2010, and till today it has hosted five rocket launchings, though not all of them successful.
5. *Delta Submarines.* Since the 1990s space object launchings have been carried out by Russian Navy Delta III and IV class submarines.[17] Usually operating from the Barents Sea Launch Area, they are reported to have conducted the first ever satellite launch from a submerged position in 1998.[18]

8.3 SWOT Analysis for Ocean-Based Space Activities

It is little known that the idea of ocean-based launchings had been studied since the dawn of the Space Era,[19] while the first ever successful launch took place just 7 years after Sputnik.[20] Following a period of hibernation, the issue surfaced again in the early 1990s, after the demise of the USSR that paved the way for private actors' involvement and eased the space race tension, and started picking up in the mid-1990s (Delta Submarines and Sea Launch). With the latest endeavor being ocean landings (SpaceX, 2015), OBSAs are now revealing their full length of capabilities.

(January–May 2015), and a totally new drone ship acquired the same name, in early 2016; see J Smedley, "SpaceX Augments and Upgrades Drone Ship Armada", online: NASA Space Flight https://www.nasaspaceflight.com/2015/06/spacex-augments-upgrades-drone-ship-armada/.

[15] The second version of "JRTI".

[16] For details see also Er Seedhouse, *Suborbital Industry at the Edge of Space*, (Switzerland, Springer International Publishing: 2014), 107–111.

[17] Delta III: Borisoglebsk, Ryazan and Delta IV: Novomoskovsk, Ekaterinburg. The first suborbital launch took place in 1995, while the first orbital launch of two nano-satellites (TUBSAT 1 and 2) was carried out by Novomoskovsk in July 1998; see Ph Clark, "Russian proposals for launching satellites from the oceans" (1999) 15 *Space Policy*, 9–12.

[18] *AU Space Primer* (Air University: Maxwell, 2003), ch. 20, p. 13, online: available at http://www.au.af.mil/au/awc/space/primer/index.htm.

[19] NASA started considering this option since 1958, when the use of USAF Texas Towers (three fixed platforms on the US East Coast) was proposed for OBSAs. Germany had also studied a similar idea for V-2 testing, in 1944; see Ch, Benson, et al., *Moonport: A History of Apollo Launch Facilities and Operations*, (NASA History Series – Special Publication 4204, 1978), online: http://www.hq.nasa.gov/office/pao/History/SP-4204/ch4-2.html#Explanation1.

[20] See OBSAs from Santa Rita Platform in 1964 (fn. 9).

OBSAs potential is rather fueled by a series of advantages, but there is also a series of factors that ought to be seriously taken into account, in order for a high degree of safety to be achieved. This part will try to lay down their main Strengths, Weaknesses, Opportunities, and Threats, not only from a technical but also from a legal and political point of view.

8.3.1 Strengths

Over 2/3 of the already carried out OBSAs were positioned near the equator, while the rest have mostly taken place around the Arctic Circle.[21] Equatorial ranges are ideal for launching Earth Observation and Telecommunications Satellites that are usually deployed in the lower Earth and geostationary orbits,[22] accordingly. Launching from the equator entails two unique physical advantages: (i) due to the Earth's shape, it provides for the shortest possible journey to equatorial and geostationary orbits, a fact that also assists in achieving biggest accuracy in orbit placement, and, most importantly, (ii) it grants maximum benefit of the so-called slingshot effect,[23] thus meaning higher velocity and quicker journey for the space object. Those two facts combined produce a series of unbeatable advantages for OBSA launchings such as (i) lesser fuel consumption that can be translated into either longer journey capability or room for larger payload and (ii) payload's operational life expansion.[24] Since launching costs can reach up to 1/3 of the overall budget[25] of a satellite lift, a significant amount of saved money can be re-routed to more substantial costs.[26]

---

[21] See Table 8.1.

[22] Out of 1738 satellites in operation, 42% are telecommunications satellites and 34% Earth observation satellites. Moreover, 61% are placed in LEO and 30% in GSO. Data retrieved from the *Union of Concerned Scientists Satellite Database*, online: UCSUSA http://www.ucsusa.org/nuclear_weapons_and_global_security/solutions/space-weapons/ucs-satellite-database.html#.VaPmmPntmkp (last accessed 15/01/2018).

[23] See "Launching Satellites," online: EUMETSAT http://www.eumetsat.int/website/home/Satellites/LaunchesandOrbits/LaunchingSatellites/index.html:

> When launching geostationary satellites, it is important that (…) [they] can be launched towards the east, where the launch impulse is aided by the spin of the Earth. This "slingshot" effect increases the speed of a launcher by 460 m/s. (…) The launch site should be as close as possible to the equator, so that the assistance is as large as possible.

[24] If larger payload volume is substituted by larger fuel load.

[25] J. J. Sellers, *Understanding Space: An Introduction to Astronautics*, (McGraw Hill Boston, 2004), 612.

[26] For example, a satellite launching into equatorial orbit from French Guyana costs 17% lower than if it had been launched from Cape Kennedy; see A Kerrest de Rozavel, "The Launch of Spacecraft from the Sea", in G Lafferranderie, D Crowther, eds, *Outlook on Space Law over the Next 30 Years* (The Hague: Kluwer International Law, 1997), 217. The importance of budgetary reductions is highlighted by SpaceX's endeavor to perform first stage landings, in order to reuse them in future operations. Contrary to the importance of equatorial position for OBSA launchings, landing positions are defined by the launching corridor used by the space object, thus meaning it can be positioned in various places around the Oceans.

Table 8.1 Technical data for OBSA platforms

Type/ Name of project	Years in action	Location/ Position	Launches/ Landings	Succ/ Fail	Source of data
Fixed Installations					
Santa Rita Platform	1964 (positioned), 1964 (all launches)	3nm off-shore Kenya (Formoza Bay)	3	3/0	*History of the Italian San Marco Equatorial Mobile Range*
San Marco Platform	1965 (positioned), 1967 (1st launch) - 1988 (last launch)	600m SE of S. Rita Platform	24	mostly successful	*History of the Italian San Marco Equatorial Mobile Range*
Floating Installations					
Sea Launch	1999 (1st launch) - still operating	International Waters about 370 km from Kiribati	36	33/ 3	http://www.sea-launch.com/
MLP Sputnik	2010 (1st launch) - still operating	within Danish Fire Practice Areas (Baltic Sea & Danish Straights)	5	Mixed results	http://copenhagensuborbitals.com/
Just Read the Instructions	2015 (all landings) - decommisioned	165 miles southeast of Charleston, S.C. (Atlantic Ocean)	2	2 Hard landings	http://www.spacex.com/
Of Course I Still Love You	2015 (1st landing)- still operating	W. Atlantic Ocean	11	8/ 3	http://www.spacex.com/
Just Read the Instructions (v. 2)	2015 (1st landing) - still operating	E. Pacific Ocean	5	4/ 1	http://www.spacex.com/
Submarines					
Delta III Submarines	1995 -still operating	Barents Sea	5	5/0	www.astronautix.com
Delta IV Submarines	1998 - still operating	Barents Sea	2	2/0	www.astronautix.com

The second and most important advantage of OBSAs are the operational and third-party safety assurances provided. Offshore activities offer better launching profiles and wider launching corridors,[27] which are not only essential for a successful operation but also eliminate possible harm in case a mission fails or is aborted after launching. The absence of human presence, man-made elements, and any other obstacles not only diminishes the possibilities of casualties and damages but also assists in optimum monitoring and tracking of the whole procedure.[28] All these lead to narrower (but targeted) protection measures and further on to lower insurance costs, since possible risks are lesser.

OBSAs are also considered to be ideal for private as well as multinational actors.[29] Moreover, launching and landing in the oceans enhance the capabilities of states that do not possess great space infrastructure or do not hold the advantage of being positioned around the equator. Hence, creating or leasing an offshore platform carrying a launch pad and positioning it in the ocean makes it easier even for non-space-faring or geographically disadvantaged[30] nations to achieve progress in space activities.

[27] For a brief analysis of the two terms, see J Carroll, *supra* note 10, 781 and 780, respectively.

[28] Direct eye contact can also reveal facts not clearly shown by telemetry.

[29] For example, this was the initial idea behind Sea Launch. On the contrary, SpaceX being involved not only in OBSAs but also in launchers' development and lifting services is trying to gain the "National Champion" position, with its budget substantially assisted by public commissions. Instead of many see Er Seedhouse, *SpaceX's Dragon: America's Next Generation Spacecraft*, (Switzerland: Springer International Publishing, 2016) 15–21.

[30] We refer to states positioned away from the equator and not to Geographically Disadvantaged States as defined in art. 70, p. 2, LOSC.

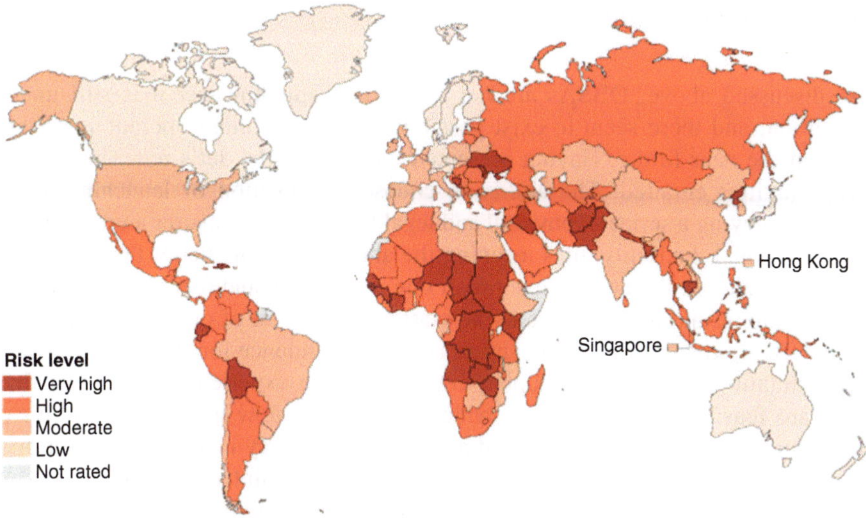

Fig. 8.1 Political instability index. (Source: Online, Economist Intelligence Unit http://graphics.eiu.com/specialReport/instability_world_map.gif)

Apart from non-space-faring nations, space powers and private actors that wish to launch or land near to the equator often try to engage in cooperative schemes with countries of the tropical zones, in order to access or create launch pads.[31] However, political–economic stability is not fundamental in those countries (see Fig. 8.1), making thus cooperation and operations even more difficult.[32] OBSAs and especially those positioned in the high seas (HS) overrun these problems. Finally, from a developing country's point of view, this type of cooperation is often criticized as New Colonialism and thus is unwanted.[33]

---

[31] Some typical examples of established launch pads are LAPAN Space Center (Indonesia), Barreira do Inferno Launch Center (Brazil), and Vikram Sarabhai Space Centre (India).

[32] See, for example, the case of massive and lengthy strikes of workers' unions in French Guiana (an Overseas Region of the French Republic) that have led into delay the launching of satellites from Kourou Spaceport in 2011 and 2017; see "Strike delays Ariane rocket launch" (online, Reuters https://www.reuters.com/article/idUSL5E7KK44020110920) and "After Strike Ends, Ariane 5 reaches French Guiana Launch Pad for Dual-Payload Delivery Mission" (online, Space Flight http://spaceflight101.com/ariane-5-va236-rollout-after-strike-delay/ (last accessed 15/01/2018)).

[33] Ronald Reagan Ballistic Missile Defense Test Site is a complex hosted in Kwajalein Atoll and Wake Island of the Republic of Marshal Islands that are leased by the US Government. Situated there for more than 40 years, it has affected RMI-US relations in several ways; see for a short comment: N Maclellan, "Kwajalein Atoll and the New Arms Race" (online: UVM http://www.uvm.edu/~jdavis6/pacific/readings/Kwajalein%2520Atoll%2520and%2520the%2520New%2520Arms%2520Race.doc). Luigi Broglio Space Centre was established in 1964 and comprised of a land-based segment on sovereign Kenyan soil and two Fixed Platforms, then located on the High Seas. Even though Kenya extended its territorial sea from 3 to 12 nm in 1972 (Art. 2, *The Territorial Waters Act*, no 2 of 1972), it was only until 1987 when the two governments signed an updated agreement regulating, among others, territory use issues; see O Ferrajolo, Launch and Tracking Stations, the "San Marco–Malindi Case," in Lafferranderie, *supra* note 26, 275.

8.3.2 Opportunities

As discussed above, OBSAs are not a new concept; however it is still under evolution, and there seem to exist possibilities for ever more operations of this kind in the near future. Having presented their Strengths, let's now shift to the Opportunities entailed. The ability to choose the position of launching, even though carrying a series of physical and legal limitations,[34] ensures a great degree of operational flexibility, thus meaning the ability to define better the launching profile, to choose the appropriate timeframe, to relocate and reprogram the mode of operations, and to store/protect the platform when out of operation.[35] In addition, OBSAs can ensure significant capacity independence for states with territorial limitations,[36] or for developing countries, since experience has shown that they are feasible even under low-budgets.[37] Turning now to legal and political considerations of OBSAs, state and non-state actors may now have a wider and more rigid variety of options as to the choice of the appointed launching state,[38,39] while at the same time, these can provide life-saving circumvention of stringent export control regimes, which may sometimes block operations on miscellaneous grounds.[40] Moreover, apart from space economy growth, OBSAs are destined to push for innovation and growth of the maritime industry as well.

[34] Such limitations will be analyzed at a later stage in this paper.

[35] It is obvious that this flexibility is more restricted for fixed platforms.

[36] For example, states distant from the equator or other preferred locations, micro-states, mountainous states, etc.

[37] See, for example, Copenhagen Suborbitals, which is a nonprofit attempt, based on voluntary work and donor funding; see "The support group" (online: Copenhagen Suborbitals http://copenhagensuborbitals.com/mission/the-support-group/, or SpaceX's re-usable rockets).

[38] The use of offshore platforms may activate the 4th tier of Article VII, OST: "each State Party from whose (…) facility an object is launched." M Sundahl discusses the options of space objects' registration launched offshore by a multinational actor; see M J Sundahl, "Legal status of spacecraft," in R S Jakhu, P S Dempsey, eds, *Routledge Handbook of Space Law*, (Oxon: Routledge, 2017), 50–52.

[39] Since space-faring nations with launching and landing capabilities are limited, so is the list of countries with appropriate National Space Legislation in force. To date, only 22 states have submitted relevant data to the UNOOSA *National Space Law Collection*; see "National Space Law Collection" (online: UNOOSA http://www.unoosa.org/oosa/en/ourwork/spacelaw/nationalspacelaw/index.html, last accessed 15/01/2018). According to von der Dunk, national space legislation does not only provide stimulus to space industry but also is of vital importance in avoiding unwanted complex legal entanglements when liability issues arise; see Fr von der Dunk, "Fundamental provisions for national space laws" in *Proceedings of the Meeting international responsibilities and addressing domestic needs*, UN, Vienna, 2006, 261–284.

[40] See C Blake, Navigating Export Controls for Small Satellites, *27th Annual AIAA/ USU Conference on Small Satellites*, 2013, for relevant difficulties that might arise.

8.3.3 Weaknesses

Operating in the marine domain does not only carry advantages, since it is considered to be a really harsh operational environment. Sea and weather conditions as well as salinity may pose significant restrictions and obstacles. From a technological and financial point of view, the development and use of state-of-the-art equipment and materials will lead to innovative space and maritime technologies. However, such development might be time and budget-consuming. This innovation nexus is also expected to involve a higher degree of risks that will surely be reflected on costs like insurance and wages.

8.3.4 Threats

Apart from typical threats generated by launching and landing activities, OBSAs can give birth to new types of threats and hazards toward inherent characteristics of the marine domain. Interference with HS freedoms or third states' rights in maritime zones,[41] disturbance of human activities due to normal operations or accidental incidents, natural environment degradation, or pollution[42] are some of the substantial dangers involved. On the legal and political side, the issue of choosing a flag of convenience for the platforms has already been highlighted,[43] while the avoidance of cooperating with equatorial developing states bans the possibility of technology and know-how transfer to those countries.[44]

8.4 Rights and Obligations of States

Launching or landing a space object is definitely a space activity, throughout the whole duration of the operation, even if it is not a success.[45] At the same time, an OBSA is also a maritime activity that has to follow specific law of the sea and

[41] See Part 4, infra.

[42] See Part 5, infra.

[43] Fr von der Dunk, "Towards 'flags of convenience' in space?", Paper Presented at the *Transfer of ownership of space objects: issues of responsibility, liability and registration* Symposium, sponsored by the International Institute of Space Law and European Centre for Space Law, 2012.

[44] Which is an established obligation according to Arts. I and X of the *Treaty on Principles Governing the Activities of States in the Exploration and Use of Outer Space, Including the Moon and Other Celestial Bodies*, 27 January 1967, 610 UNTS 205, 18 UST 2410, TIAS No 6347, 6 ILM 386 (entered into force on 10 October 1967) (hereafter "Outer Space Treaty") and Art. 4 of the *Agreement governing the Activities of States on the Moon and Other Celestial Bodies*, 5 December 1979, 1363 UNTS 3 (entered into force 11 July 1984) (hereafter "Moon Agreement").

[45] Article II of the *Convention on International Liability for Damage Caused by Space Objects*, 29 March 1972, 961 UNTS 187, 24 UST 2389, 10 ILM 965 (1971) (entered into force 1 September

maritime law regulations. Based on a general law of the sea rule which dictates that "land dominates the sea,"[46] we will examine the deployment of OBSAs according to the position and space they occupy, in order to analyze the different sets of rights and obligations for either Launching States or third States.

The law of the sea portions the oceans into different types of zones, according to the degree of state sovereignty appointed to them, which may vary from absolute to zero. For the scope of this paper, we will divide the aforementioned zones into three distinct categories, and further on we will examine the ability of Coastal States to regulate and conduct OBSAs within them.

8.4.1 *Zones of Strict Control (Areas of National Sovereignty)*

Internal waters (IW) occupy "the landward side of the baseline of the territorial sea."[47] Usually they take the form of bays, estuaries, ports, etc., and their legal status is assimilated to the one of land territory,[48] which means total and unrestricted control by the Coastal State for every use. Thus OBSAs can take place only under its consent and authorization, albeit a very rare fact to occur, since IW are considered to be critical not only for national security but also for various commercial and public activities, or may have significant ecological values.

The territorial sea (TS) is considered to be the state's spatial expansion into the marine domain. Within a zone of up to 12 nm from the baseline, the Coastal State exercises absolute and exclusive sovereignty,[49] to the exception of innocent passage.[50] This right is granted to third states but only for navigational purposes, while all other uses are excluded. Article 19 explicitly refers to a ban of "(e) the launching, landing or taking on board of any aircraft; (f) the launching, landing or taking on board of any military device; [and] (l) any other activity not having a direct bearing on passage," while submarines traversing the TS "are required to navigate on the surface and to show their flag."[51] The Coastal State also withholds the right

1972) (hereafter "Liability Convention") states that "A launching State shall be absolutely liable to pay compensation for damage caused by its space object on the surface of the Earth or to aircraft in flight" (similar wording is used in Art. VII of the Outer Space Treaty).

[46] North Sea Continental Shelf (Federal Republic of Germany v Denmark) (Merits) (1969) ICJ Rep 3 para. 96. For detailed analysis see Ir Papanicolopulu, "The land dominates the sea (dominates the land dominates the sea)" *Questions of International Law*, 47 (2018).

[47] See Art. 8 LOSC.

[48] A relevant example of IW regime application is found in Kennedy Space Centre, Cape Canaveral. Even though launch pads are positioned on reclaimed land (not platforms), they are regulated by IW legislation. See 33 CFR 165.701, Vicinity, Kennedy Space Center, Merritt Island, Florida, security zone.

[49] See Art. 2, LOSC.

[50] See Art. 17, LOSC.

[51] See Art. 20, LOSC.

to create or authorize any fixed platform and its operation,[52] and also holds the exclusive right to protect it from any threat.[53]

In the case of archipelagic states, the Archipelagic Waters (AW)[54] regime applies, within which full sovereignty applies in a way similar to the TS.[55] According to LOSC, the right of innocent passage is also valid in AW,[56] and with it come all the rules spelled out in Part II of the Convention.

Coastal States have strict control over areas of national sovereignty and hold the exclusive power to authorize, regulate, and ban OBSAs for nationals and non-nationals.

8.4.2 *Zones of Use-Related Control*

To the exception of the continental shelf (CS), which exists ipso facto and ab initio for every Coastal State, the contiguous and the Exclusive Economic Zones (EEZ) are optional. In the contiguous zone[57] (CZ), the Coastal State has only police and repression authority over violations of customs, fiscal, immigration, and sanitary national regulations that are about to occur or have already occurred within its territory or TS, respectively. Apart from this authority, and unless an EEZ exists, the CZ is considered to be part of the HS.[58]

The CS comprises only the seabed and subsoil of an area adjacent to the TS and normally reaches up to 200 nm from the baseline,[59] over which the Coastal State has natural resources exploration and exploitation sovereign rights. Even though OBSAs do not fall under this category,[60] Coastal States "have the exclusive right to construct

[52] A right granted by the full sovereignty status.

[53] Coastal State may adopt regulations on "the protection of (...) facilities or installations" (Art. 21. b), while "any act aimed at interfering with any systems of communication or any other facilities or installations of the coastal State" is also considered to be a violation of Innocent Passage (Art. 19. k).

[54] Archipelagic Waters consist of the area included within and after the drawing of the archipelagic baselines (Art. 47, LOSC) and contain the sea, the sea floor, as well as the air space above them. Apart from them, archipelagic states are entitled to all other types of Zones (Art. 48, LOSC).

[55] See Art. 49, para. 1, LOSC.

[56] See Art. 52, LOSC.

[57] Maximum breadth of 24 nm from the baselines, Art. 33, LOSC.

[58] For relative analysis, see infra. The CZ repression authority applies on internationally wrongful acts always in connection to the land or the TS of a coastal state, while the airspace above it is international. Hence it is almost impossible for an OBSA taking place in a CZ to pose threat to any related customs, fiscal, or immigration rules. However, in case an object landing goes wrong and there is possibility to enter/or if it enters the TS, then sanitary regulations against backward contamination (Art. IX, OST) might be applicable, depending on national legislation.

[59] Although exceptions do apply under conditions described in Art. 76, LOSC.

[60] We are referring to launch pads carried on fixed/floating platforms and not on submarines.

and to authorize and regulate the construction, operation and use of: (…) installations and structures which may interfere with the exercise of the rights of the coastal State in the zone,"[61] thus gaining a superior position over any kind of installation aiming at being positioned on (fixed) or above (floating) its CS, and especially for fixed platforms and anchored floating platforms, it has "exclusive jurisdiction (…) including jurisdiction with regard to customs, fiscal, health, safety and immigration laws and regulations."[62]

The same status applies on the EEZ, which extends up to 200 nm from the baseline,[63] and apart from the seabed rights, it also entails fisheries rights, renewable energy production in the water column, and marine scientific research jurisdiction.[64] Moreover, the Coastal State has jurisdiction with regard to the protection and preservation of the marine environment,[65] enhancing thus their environmental regulatory authority over OBSAs. However, it is clearly stated that the "Coastal State shall have due regard to the rights and duties of other States" which still possess the right to access and use of the EEZ for all other non-exclusive uses.

In both economic zones, authorizing states have the right of establishing a safety zone in order to ensure the safety of navigation as well as of any activity carried from a platform, which shall not exceed the distance of 500 m.[66]

Coastal States impose specific authorities on the economic interest zones and hold a clear priority of regulation and use over them. However they must show "due regard" for any other use or user, since they only hold exclusive rights of natural resources exploration and exploitation, but do not exercise sovereignty on those zones (EEZ and CS). The vastness of those areas and their relatively small depth and secure distance from near-shore areas and activities make them ideal for OBSAs. At the same time, Coastal States' rights ought to be safeguarded, and thus possible conflicts might arise.

---

[61] See Art. 60.1.c, LOSC.

[62] See Art. 60.2, LOSC.

[63] See Art. 56, 57, LOSC.

[64] See Art. 56, LOSC. For all other uses (e.g., submarine operations), the water column of the EEZ is considered to be assimilated to the high seas; see Art. 58, para. 1, LOSC.

[65] Art. 56, para. 1, LOSC.

[66] See Art. 60, paras 4–7, LOSC. The breadth of this zone has often been criticized as insufficient, and several states have asked for its enlargement; for details see A Harel, Preventing Terrorist Attacks on Offshore Platforms: Do States Have Sufficient Legal Tools? (2012) 4:1 *Harvard National Security Journal*, 131–184. In a similar vein, Laubscher and Nase have suggested for a zone of not less than 40 km of sea surface and 100 km of air space around space elevator facilities, in order for timely warning to be provided to nearby sea and air navigation; see R. S. Jakhu, J N Pelton, eds, *Global Space Governance: An International Study* (Switzerland: Springer International Publishing AG, 2017) 242.

8.4.3 Areas of Common Use

The marine domain that falls outside national jurisdiction can be divided into two distinct areas: (a) the HS that comprises only of the water column not included in the EEZ, in the TS, or in the IW of a State or in the AW of an archipelagic state[67] and (b) The Area, which comprises of the seabed and ocean floor and subsoil thereof beyond the limits of the previously mentioned zones, the CS also excluded.[68] The main difference between those two is that the HS is a domain of common use, while The Area is a domain of collective management; however both of them are open for access and use and cannot be subjected to any kind of state sovereignty.

Use of the HS is principled by the so-called Freedoms of the HS, which are non-exhaustibly listed in the LOSC.[69] Even though launching and landing of space objects are not explicitly referred, they can be exercised as long as they are compatible to LOSC regulations, and they are exercised with due regard for interests of other states in the HS and The Area[70] and serve peaceful purposes.[71] So, in the absence of any type of state sovereignty, the responsibility for authorization, regulation, and enforcement now falls into the hands of the flag state of any OBSA-engaged platform.[72]

The Area is a sui generis legal construction that bears great resemblance to celestial bodies, as it has, along with its resources, been declared as common heritage of mankind,[73] and it cannot also be subjected to any kind of national sovereignty.[74] Contrary to the longstanding and customary rights of the HS, this relevantly new regime constrained LOSC authors from laying down rights and obligations in a detailed way.[75] Consequently, OBSAs are permitted as long as they

---

[67] See Art. 86, LOSC.

[68] See Art. 1, LOSC.

[69] These are navigation, fishing, scientific research, overflight, laying of submarine cables and pipelines, and construction of artificial islands and other installations (Art. 87, LOSC), with the last three being essential to OBSAs.

[70] See Art. 87, LOSC.

[71] See Art. 88, LOSC. Even though there exists a Peaceful Uses sweeping clause in Art. 301 of the convention, authors felt the necessity of adding a Peaceful Purpose clause within Part VII and many more in Part XI – The Area.

[72] Flag state and launching state may not always coincide, making thus authorization issues more complex. See, for example, the Sea Launch case, analyzed in von der Dunk, Fr., Sovereignty Versus Space – Public Law and Private Launch in the Asian Context, *Space and Telecommunications Law Program Faculty Publications*, Paper 1, 2001, pp. 42–44.

[73] See Art. 136, LOSC.

[74] See Art. 137, para. 1, LOSC.

[75] This hesitation is reflected in Art. 147, where paragraph 1 writes "Activities in the Area shall be carried out with reasonable regard for other activities in the marine environment" vis-a-vis paragraph 3 "Other activities in the marine environment shall be conducted with reasonable regard for activities in the Area," in an attempt to achieve a balance between uses.

are in conformity with Part XI–LOSC and the UN Charter[76] and do not aim at appropriating any part of The Area[77] or using it for non-peaceful purposes.[78] Responsibility for OBSAs authorization, regulation, and enforcement is identical to the one of the HS.[79]

The legal regime laid down for the Common Use areas is based on the HS Freedoms on the one hand and on the other on nationality or flag state jurisdiction over crafts and installations. These two are also accompanied by third-state intervention rights in specific cases of universal or international crimes[80] and the general non-harm of areas outside national jurisdiction rules[81] established in space law,[82] environmental law,[83] and law of the sea.[84]

8.5 Environmental and Biodiversity Challenges

Apart from the typical externalities that land-based launching and landing activities entail, OSBAs are believed to carry many more due to the different nature and vulnerability of the marine domain. Even though there is a large amount of knowledge and regulations concerning the environmental parameters of Artificial Islands and Structures, mainly related to oil and gas exploitation,[85] there seems to be a gap of relevant knowledge for installations deployed in OBSAs.

Environmental consequences can have short-term (pollution), medium-term (contamination), and long-term effects (degradation), according to the type and volume of input. Based on oil and gas industry experience, environmental threats may occur during the three phases of the operational life of an installation, namely, during construction, operation, and decommission. During the construction of fixed installations (or anchoring – for floating), phase-specific care should be taken in order to avoid the short- or long-term destruction of fragile ecosystems and

[76] See Art. 138, LOSC.

[77] See Art. 137, LOSC.

[78] See Art. 141, LOSC.

[79] Especially referring to fixed or anchored floating installations.

[80] See Art. 110. LOSC. For a short description, see G. Tsaltas and G. Rodotheatos, "Maritime Interdiction Operations: A View through International Law Lens" (2010) 2 *NMIOTC MIO Journal*, 45–48.

[81] For a general comment, see P. Manikowski, "Examples of space damages in the light of international space law" (2006) 6:1 *Poznan University of Economics Review*, 54–68.

[82] See Art. 2 of the Liability Convention, Art. 9 of the Outer Space Treaty, and Art 7 of the Moon Agreement.

[83] See Principles 2 and 13, Rio Declaration on Environment and Development, 1992.

[84] See Art. 194, para. 2, LOSC.

[85] Instead of many, see J. Rochette et al., *Seeing beyond the horizon for deepwater oil and gas: strengthening the international regulation of offshore exploration and exploitation*, IDDRI Study, no 1/14, February 2014.

habitats, because of dredging, explosions, or introduction of invasive species.[86] The potentially most dangerous phase is the second one; threats such as dumping, noise pollution, and thermal pollution may have devastating effects for biodiversity and the environment. Finally, the seemingly innocent phase of decommission carries a series of environmental and navigational threats if it is not planned and executed cautiously.[87]

The previously mentioned threats concern not only launchings but also landings. However, for the latter case, extra precautions should be taken in order to prevent backward contamination.

8.6 Discussion

Even if the amount of OBSAs is relatively small compared to land-based launchings and landings, it is indicative of their feasibility. The SWOT analysis pointed out a series of important strengths that are in place, and interesting opportunities that should be sought after. Apart from the typical actors (i.e., states and public authorities), OBSAs seem to be ideal for private actors, research and academia, NewSpace initiatives, and developing states, which are the ones currently leading this endeavor. However, if progress is to be achieved, innovation and impacts will spread across the space sector, as well as the maritime sector.

The interface between space law and law of the sea, that even though share much in common, is minimal and generic, since there are no specific cross-referring regulations in both branches. Not surprisingly, until today the simple and small amount of OBSAs has not provoked any large conflicts between space and maritime activities. However, the ever expanding use of the marine domain, if combined with an increase of space activities,[88] may push things to the adoption of new rules concerning prioritization of uses, regulation of liability issues, and environmental protection.

8.7 Conclusion

Having identified that the way to move space launching and landing activities toward the sea exists, it remains to be seen if there is a will. This situation can be compared, up to a point, with the construction of airports on artificial islands and

[86] The problem is more likely to spread with the deployment of floating installations and submarines that may operate in different spots.

[87] For experience gained from the Oil and Gas Industry, see P. Ekins et al., "Decommissioning of offshore oil and gas facilities: a comparative assessment of different scenarios" (2006) 79:4 *Journal of Environmental Management*, 420–438.

[88] The dynamics of the Space Industry is described in OECD, *The Space Economy at a Glance 2014*, OECD Publishing, 2014.

reclaimed land, a practice mainly implemented in Asia.[89] Civil aviation seems to have grasped the various advantages entailed in this offshore activity, and as experience has shown, it has been a successful endeavor, at least from an operational point of view.[90]

In a similar way, OBSAs are ideal for specific uses and purposes (e.g., launching from the equator, or offshore landings), but it is highly impossible that they can substitute all land-based activities. For this reason, stakeholders and regulatory authorities should focus on specific uses and work together in order to create a coherent framework, which would pave the way for more new activities. Special consideration should be given to the HS and The Area, whose public and open profile should be preserved. To this aim a set of relevant regulations for OBSAs could be drafted through the competent authorities, such as COPUOS (general framework), IMO (pollution issues), and Meetings of States Parties to the LOSC (environment and biodiversity protection).

[89] Where exists a series of airports already operating: Kansai, Kobe, Kitakyushu, Chūbu Centrair (Japan), Chep Lap Kok (Hong Kong), Macao, Incheon (S. Korea), and Ordu-Giresun (Turkey). There are also scenarios for Europe (The Netherlands has considered the option of constructing an airport in its EEZ). For details on the legal ramifications of offshore airports, see P M de Leon and E J Molenaar, "Still a Mile too Far? International Law Implications of the Location of an Airport in the Sea", (2004) 14:1 *Leiden Journal of International Law*, 233–245.

[90] It has been observed that offshore airports are suitable for specific situations and are not considered to be a generic solution; see P Nijkamp and H Yim, "Critical Success Factors for Offshore Airports. A Comparative Evaluation", (2000) 35 *Serie Research Memoranda*, no 35, Faculteit der Economische Wetenschappen en Econometric, Vrije Universiteit Amsterdam.

Chapter 9
Judicial Settlement of Space-Related Disputes: Sovereignty's Final Fetters

Stratis G. Georgilas

9.1 Introduction

International space law has met or is about to meet its maturity. Still, being presumably the youngest subfield in the discourse, it appears that both theory and practice were preoccupied with the acute development of its substantive – as opposed to its procedural – part. Quite rightly, too, for technology and indeed space technology has advanced tremendously during the last 70 years. It was an unprecedented leap.

The overall situation forced laymen around the globe to breathlessly follow new, powerful and innovative techniques while struggling to clarify this ex post consensual legal regime. In a much celebrated phrase originating in 1966, the legal profession tried its utmost to 'boldly go where no one has gone before'.

Notwithstanding the Cold War at its apogee, States successfully concluded what we nowadays refer to as the *corpus* of international space law. Alas, the absolute concept of sovereignty and, to put it mildly, the scepticism about the neutrality of men and institutions, the latter duly established and mandated to resolve international disputes, was a quite known secret. The 'socialist bloc', as it then was,

Barrister, LL.M. (i) (*Cantab.*, *Hughes Hall*), *G – H Chambers*, Head of Chambers, Of the Athens' Bar, www.ghlaw.gr. 2015 ©.

I am indeed indebted to *J. Crawford* (*Cantab. & I.C.J.*), *Chr. Greenwood* (*Cantab. & I.C.J.*) and *V. Lowe* (*Cantab. & Oxon.*). The usual disclaimer applies.

S. G. Georgilas (✉)
Georgilas–Hatzopoulos Law Chambers, Athens, Greece
e-mail: egeorgil@otenet.gr; stratis@ghlaw.gr

G. D. Kyriakopoulos, M. Manoli (eds.), *The Space Treaties at Crossroads*,
https://doi.org/10.1007/978-3-030-01479-7_9

continued its general practice and 'adopted a systematic attitude [...] of refusal to submit to the jurisdiction of the Court in general'.[1]

The disaster of 1966 in the *South West Africa case* offered novel ammunition to the magazine of this school of thought and State practice.[2] Its paramount effect on many States' consciousness that current interests could be transformed into the future – often relentless conflicts – credited substantially their scepticism in submitting themselves to any third-party dispute settlement mechanism. To one's detriment, so much have minds become preoccupied with the idea of changing the law (space law included) that, in the prophetic words of Hersch Lauterpacht, writing in 1937, 'they have become insensible to the benefits of its normal observance and ascertainment'.[3]

This paper suggests that the principal judicial organ of the UN[4] may, nonetheless, adjudicate successfully, today and as the law stands, disputes deriving from international space law. Before doing so, I intend to briefly present, arbitrarily no doubt, certain issues on judicial settlement of international disputes. Treaties for the peaceful resolution of disputes, special agreements and *forum prorogatum* as jurisdictional bases of the Court will not be addressed in this paper. Neither will this paper cover disputes' and claims' justiciability (or non), whether they originate from space law or not. Despite the foregoing, an admission must be entered, namely, all cases brought before the Court have been of utterly political character, one way or another. Practice has provided ample evidence that each international situation is

[1] G Abi-Saab, The International Court as a World Court in V Lowe & M Fitzmaurice (Eds.), 50 years of the International Court of Justice (CUP 1996) 3, 4. The political consideration was that State sovereignty could be impinged upon due to the compulsory jurisdiction of the World Court (both the Permanent Court for International Justice and the International Court of Justice, henceforth, the Court). Allot puts forward five arguments supporting his view that it is necessary to question whether the Court is a court of law, let alone a court of justice, P Allot, The International Court and the Voice of Justice in V Lowe & M Fitzmaurice (Eds.), ibid., 27. Kooijmans observes that the Court is to a large extent part of the Westphalian system and, at the same time, being one of the main organs of the UN, an organization designed as an instrument to break away from the Westphalian order; however, hesitantly and haltingly, this was put into practice, P Kooijmans, The International Court of Justice: Where does it Stand? In A. Müller, D Raič & J Thuranszky (Eds.) The International Court of Justice – Its Future Role after 50 Years (Martinus Nijhoff 1997) 407.

[2] *Second Phase (Ethiopia & Liberia v. South Africa), Judgment of 18 July 1966, I.C.J. Reports (1966)*, at 6. Abi-Saab suggests that a new self-awareness and change of attitude on the part of the Court itself were presumably the most significant consequence of this crisis (note 1) 6.

[3] H Lauterpacht, Peaceful Change – The Legal Aspect in Sir Elihu Lauterpacht (Ed.), Hersch Lauterpacht, International Law – Collected Papers, Vol. 5, Disputes, War and Neutrality (CUP 2004) 10.

[4] M Sameh Amir, The Role of the International Court of Justice as the Principal Judicial Organ of the United Nations (Martinus Nijhoff 2003). Bowett and Others, The International Court of Justice: Process, Practice and Procedure (B.I.I.C.L. 1997). For a recent, concise and updated review of the Court's activities S Murphy, The International Court of Justice in C Giorgetti (Ed.) The Rules, Practice and Jurisprudence of International Courts and Tribunals (Martinus Nijhoff 2012) 11.

capable of being determined as a matter of law.[5] Lastly, as far as space-related disputes between commercial entities and States are concerned, well-documented proposals have been already published. With respect, the particular type of disputes is well beyond the ambit of this presentation.

In the context under scrutiny, one should be always mindful that 'the rule of law is a rule of integrity'[6] and 'one of the most powerful words or expressions today'.[7] In Tom Bingham's words, 'observance of the rule of law is quite as important on the international plane as on the national, perhaps even more so',[8] for it mirrors the history of efforts to restrain unchecked sovereign power.[9] Following 1928 and, by all means, 1945, members of the international community were keen on submitting their disputes to a binding decision of a *forum* adjudicating upon the basis of the law. This duty was considered to be 'the first and primary condition of a political society worthy of that name'.[10]

Against this background this paper will be devoted to what the author considers as an essential admission for resolving space-related international disputes of today and tomorrow.

9.2 Certain Issues on Judicially Settling International Disputes

The international community rejected serious proposals for the compulsory jurisdiction of international disputes three times during the twentieth century, namely, in 1907, 1920 and 1945.[11] Currently, sovereign independence of States is indeed curtailed by the international rule of law. Its manifestation is twofold.

---

[5] Oppenheim's International Law by Sir Robert Jennings & Sir Arthur Watts (9th Edition, Longman 1992) 12–13.

[6] Lord Neuberger MR, 'The Tyranny of the Consumer and the Rule of Law' (25th Annual Bar Conference, 6 Nov 2010), in file.

[7] B Tamanaha, On the Rule of Law (CUP 2004) 3.

[8] T Bingham, The Rule of Law (Allen Lane 2010) 110.

[9] N Politis, 'Le Problème des Limitations de la Souveraineté et la Theorie de l' Abus des Droits dans les Rapports Internationaux', (1925) 1 RC 5. S. Chesterman, 'An International Rule of Law?', available at http://ssrn.com/abstract=1081738 (last visit 30 Jul 2015) 38. J Morin spoke of an emerging international rule of law, 'L' État de Droit: Emergence d' un Principe du Droit International' (1995) 255 RC 9, 27. I Brownlie, The Rule of Law in International Affairs (Brill 1998). For a quite novel approach, E Benvenisti, 'Sovereigns as Trustees of Humanity: On the Accountability of States to foreign Stakeholders' (2013) 107 AJIL 295.

[10] H Lauterpacht, The Judicial Settlement of International Disputes in Sir Elihu Lauterpacht (Ed.), (note 3) 72.

[11] S Rosenne, Three Central Elements of Modern International Law in Essays on International Law and Practice (Martinus Nijhoff 2007) 621. As per the concepts of statehood and sovereignty, inter alia, J Crawford, The Creation of States in International Law (2nd Edition, OUP 2007) 31. The Court has stated that '[I]t is sufficient to say that State sovereignty evidently extends to the area of its external policy, and that there is no rule of customary international law to prevent a State from

Since 1945 no State has ever suggested that it is legally admissible to consider war – or better armed conflict – as an institution for changing the law. As Judge Sir Christopher Greenwood put it, '[…] War is the antithesis of the rule of law. […] The reality is that the rule of law in international society cannot be separated from the rule of law within the state […] To me, respect of the rule of law is fundamental for preserving peace'.[12]

Secondly, States accepted the legal duty to acquiesce in changes in the law decreed by a competent international organ.[13] The existence of such duty toppled with the obligation of States to settle their disputes amicably and in accordance with international law, if possible administered by a Court,[14] constitutes a basic feature of the UN Charter. State practice reveals that most, if not all, judgements rendered by the Court are accepted as influential contributions to the legal process[15] and the pacification of international society.[16] So far, its *dicta* managed to balance successfully third-party interests; instituting proceedings is not a rare phenomenon anymore.

choosing and conducting a foreign policy in co-ordination with that of another State', *Military and Paramilitary Activities in and against Nicaragua (Nicaragua v. USA), Merits, Judgment of 27 July 1986, I.C.J. Reports (1986) p. 133 para. 265*, also available at http://www.icj-cij.org/docket/files/70/6503.pdf (last visit 30 Jul 2015). It had already observed that 'it is no doubt true that the Islamic Revolution of Iran is a matter "essentially and directly within" the national sovereignty of Iran; […] however a dispute which concerns diplomatic and consular premises and the detention of internationally protected persons, and involves the interpretation or application of multilateral conventions codifying the international law governing diplomatic and consular relations, is one which by its very nature falls within international jurisdiction', *United States Diplomatic and Consular Staff in Tehran (USA v. Iran), Provisional Measures, Order of 15 December 1979, I.C.J. Reports (1979) p. 15–16 para. 25*, also available at http://www.icj-cij.org/docket/files/64/6283.pdf (last visit 30 Jul 2015). As per the so-called Treaties of Protection and their impact on sovereignty, the Court held that they amount to 'a form of organization of a colonial territory on the basis of autonomy of the natives [and] suzerainty over the native States becomes the basis of territorial sovereignty as towards other members of the community of nations', *Land and Maritime Boundary between Cameroon and Nigeria (Cameroon v. Nigeria: Equatorial Guinea intervening), Merits, Judgment of 10 October 2002, I.C.J. Reports (2002) p. 405 para. 205*, also available at http://www.icj-cij.org/docket/files/94/7453.pdf (last visit 30 Jul 2015).

[12] The Rule of Law in International Society (The First Milon Kumar Banerji Memorial Lecture 16 Jan 2013), available at http://www.youtube.com/watch?v=JNaup3oQzxM (last visit 30 Jul 2015).

[13] All members of the UN are ipso facto contracting parties to the Court's Statute and (are) bound by the latter's pronouncements, not in the sense of *res judicata* but as authoritative findings as per the substantive rules and norms of international law.

[14] It must be remembered that in February 1944 the Inter-Allied Committee on the future of the PCIJ adopted a report. It was proposed that the revised Statute of the Court should not provide for any kind of obligatory jurisdiction, for any such step was considered premature. Inter-temporal issues on the jurisdiction of the Court are well addressed in N Gallus, The Temporal Jurisdiction of International Tribunals (OUP 2017).

[15] C Tams, The ICJ as a 'Law-Formative Agency': Summary and Synthesis in C Tams & J Sloan (Eds.) The Development of International Law by the International Court of Justice (OUP 2013) 379.

[16] P Couvreur, The Effectiveness of the International Court of Justice in the Peaceful Settlement of International Disputes in A. Müller, D Raič & J Thuranszky (Eds.) (note 1) 85.

After all, the Court enjoys a potentially unlimited scope of jurisdiction *ratione materiae*. Judge Sir Robert Jennings was mindful that

> resort to third party settlement by definition transfers to others the direct responsibility. In a word, there is somebody else to blame. But there is another reason why the Court is good for hard cases: I refer to a quality that is part and parcel of the process of adjudication by a Court of Law.[17]

States frequently hesitate to refer some disputes to the Court, for they fear the solution to be rendered cannot be kept within their control.[18] Had 'State x' perceived its claim utterly superior than its adversary's 'State z', the former would surely opt for judicial settlement. In principle, adjudication applies legal criteria. The question of how far alternative means can be regarded as a satisfactory substitute for obligatory judicial settlement of disputes is an old one. Indeed, in 1930 it was maintained that:

> [W]hatever the advantages of conciliation may be, it is clear that [...] it does not aim at securing peace within the international society by ascertaining in a final and binding manner the disputed rights. [The judicial settlement's] primary task, a function which is essential to the existence of a community under the reign of law [...] The moral and political effect of the decision of a judicial tribunal is so tremendous that its legal finality has a tendency to approach actual finality. [...] The absence of the obligation to submit to a final judicial pronouncement [...] amounts in practice to enabling the state, which has committed a wrong or is in wrongful possession to defy the law without the illegality of its conduct being impartially ascertained and this in fact alone may prove a menace to peace.[19]

No State has ever disputed the usefulness of the so-called political means in peacefully adjusting international controversies. It involves 'criteria derived from considerations of politics, statesmanship, convenience, expedience and so on. The possible variants are endless'.[20]

The two methods are widely perceived as complimentary to one another, without the choice of one necessitating the rejection of the other.[21] Presumably, the core questions are how far States accept the need for judicial settlement and why recourse to the Court is still regarded as a rather exceptional method of resolving international disputes?

---

[17] Sir Robert Jennings, Contributions of the Court to the Resolution of International Tensions in C. Peck & R. Lee (Eds.) Increasing the Effectiveness of the International Court of Justice (Martinus Nijhoff 1997) 83.

[18] M Bedjaoui, The International Court of Justice at 50 – Introduction in C. Peck & R. Lee (Eds.) (note 17) 20. Also, A Adede, Judicial Settlement in Perspective in A. Müller, D Raič & J Thuranszky (Eds.) (note 1) 49, V Lowe, 'The Function of Litigation in International Society' (2012) 61 ICLQ 209.

[19] H Lauterpacht, The Absence of an International Legislature and the Compulsory Jurisdiction of International Tribunals in Sir Elihu Lauterpacht (Ed.), (note 3) 205–206, emphasis added.

[20] Sir Robert Jennings, The proper Work and Purposes of the International Court of Justice in A. Müller, D Raič & J Thuranszky (Eds.) (note 1) 37.

[21] G Guillaume, 'The Future of International Judicial Institutions' (1995) 44 ICLQ 848.

9.3 Judicial Settlement of 'Space-Related' Disputes by the Court

State practice reveals an excellent track of compliance with established space law principles.[22] As already indicated, this formidable body of law was created in a relatively short period.

There is absolutely no need to spend more lines stating the obvious: States that participated in the final drafting of the 'space treaties' did not reach an agreement tackling the need of a conventional type compulsory judicial settlement mechanism, that is, a compromissory clause.[23] As a general term of reference, this legal sub-regime contains little provision with respect to the resolution of disputes. The idea of settling judicially a space dispute was not treated as a panacea that will secure international peace and solve international conflicts.

State practice confirms that sovereignty, as far as space disputes are concerned, remained relatively free of any relevant conventional impingement. This practice vindicated sovereignty (and indeed the fullest international capacity)[24] enjoyed by the States. Of importance it is also relevant to remember that, during the late 1960s and 1970s, many States announced a policy of denial to acknowledge customary international law bequeathed by the nineteenth-century Europe.[25] Some States reverted to old techniques of dispute settlement.[26] Arbitration of space law disputes has been the preferred tool, albeit restricted to instruments outside the UN framework.

---

[22] F Lyall & P Larsen, Space Law – A Treatise (Ashgate 2009) 559.

[23] With regard to the Outer Space Treaty, the USSR insisted on 'consultation' (UN Docs A/6352 of 16 Jun 1966), while the USA supported a 'referral clause' to the Court (UN Docs A/AC.105/32 of 17 Jun 1966).

[24] It has been pronounced that '[t]he subjects of law in any legal system are not necessarily identical in their nature or in the extend of their legal rights, and their nature depends upon the need of the community', *Reparation for Injuries Suffered in the Service of the United Nations, Advisory Opinion of 11 April 1949, I.C.J. Reports (1949), p. 178*, also available at http://www.icj-cij.org/docket/files/4/1835.pdf (last visit 30 Jul 2015). The legal nature and status of a non-State entity, albeit in a limited context, was examined by the Court, its Opinion being that 'taking into account the fact that the General Assembly has granted Palestine a special status of observer and that the latter is co-sponsor of the draft resolution requesting the advisory opinion [...] Decides [...] Palestine may also take part in the hearings', *Legal Consequences of the Construction of a Wall in the Occupied Palestinian Territory, Order of 19 December 2003, I.C.J. Reports (2003), p. 429, para.* 2, also available at http://www.icj-cij.org/docket/files/131/1527.pdf (last visit 30 Jul 2015).

[25] S Rosenne, The Role of Controversy in International Legal Development in (note 11) 561. It encapsulated the traditional belief that 'any surrender of the powers associated with sovereignty is an acknowledgement of national weakness and will indeed lead to a prerogative diminution of state power', analysed by C Weeramantry, Emerging Dimensions of Sovereignty in International Law, in Universalising International Law (Martinus Nijhoff 2004) 106.

[26] G Meishan-Goh, Dispute Settlement in International Space Law (Martinus Nijhoff 2007) 22.

The 1972 Convention on International Liability for Damage Caused by Space Objects[27] remains the sole exception. A specific Claims Commission was established for examining any claim for compensation, originating from damage caused by the said objects. This body follows closely the well-established pattern without any deviation,[28] but the discussion of technical details does not fall within the scope of this paper.

While it seems difficult to identify areas of international law in which the Court's holdings are of no relevance, its footprint is not quite visible in space-related disputes. This does not diminish its overall impact. Sir Franklin Berman quite recently remarked that:

> the 'founding intent' remains intact, namely to have an institution of the same essential kind as the one conceived in 1920 [...] this was to be an institution for settling disputes, and by doing so it was to contribute to world peace. Any development of the substantive system of the law would follow as a by-product of that process, but not as an essential aim in its own right [...][29]

the comment being in complete harmony with the Court's findings some 40 years ago, when it held that 'the Court as a court of law cannot render judgment *sub specie legis ferendae*, or anticipate the law before the legislator has laid it down'.[30]

9.4 The 'Optional Clause' Revisited: Are 'Space-Related' Disputes Part of the Court's Jurisdiction?

Some 15 years ago, Judge Oda observed that '[t]he idea that the optional clause should be accepted by all States with the least possible reservation and without any fixed period of validity seems to be a fairly popular notion'.[31]

---

[27] *Convention on International Liability for Damage Caused by Space Objects*, 29 March 1972, 961 UNTS 187, 24UST 2389, 10 ILM 965 (1971) (entered into force 1 September 1972) [Liability Convention].

[28] L Brillmayer (et al.), International Claims Commissions: Righting Wrongs after Conflict (Elgar 2017).

[29] Sir Franklin Berman, The International Court of Justice as an 'Agent' of Legal Development in C Tams & J Sloan (Eds.) (note 15) 10.

[30] *Icelandic Fisheries Case (UK v. Iceland), Merits, Judgment of 25 July 1974, I.C.J. Rep. (1974) p. 23–24, para. 53*, also available at http://www.icj-cij.org/docket/files/55/5977.pdf(last visit 30 Jul 2015).

[31] S Oda, 'The Compulsory Jurisdiction of the International Court of Justice' (2000) 49 ICLQ 251. Also S Rosenne, The Law and Practice of the International Court, 1920–2005, Volume II (Martinus Nijhoff 2006) 701, J Merrills, The Optional Clause at Eighty in N. Ando, E McWhinney & R Wolfrum (Eds.) *Liber Amicorum Judge Shigeru Oda* (Brill Nijhoff 2002), vol. 1, 438, R Kolb, The International Court of Justice (Hart 2013) 447, J Quintana, Litigation at the International Court of Justice (Brill Nijhoff 2015) 95 and C Tomuschat, in A Zimmermann et al., The Statute of the International Court of Justice (OUP 2006) 589.

The Court repeatedly pronounced that its jurisdiction is based on both Article 36 § 2 of its Statute and the submitted unilateral optional clause declarations. It emphasized that it enjoys its jurisdiction 'only as far as the two unilateral declarations coincide'[32] and that:

> declarations, even though they are unilateral acts, establish a series of bilateral engagements with other States accepting the same obligation of compulsory jurisdiction, in which the conditions, reservations and time limit clauses are taken into consideration. In the establishment of this network of engagements, which constitutes the Optional-Clause system, the principle of good faith plays an important role.[33]

It also clarified that

> any State party to the Statute, in adhering to the jurisdiction of the Court in accordance with Article 36, paragraph 2, accepts jurisdiction in its relationship with States previously having adhered to that clause. At the same time, it makes a standing offer to the other States party to the Statute, which have not yet deposited a declaration of acceptance. The day one of those States accepts that offer by depositing in its turn its declaration of acceptance, the consensual bond is established and no further condition needs to be fulfilled.[34]

Especially, as per the interpretation of declarations made under article 36, paragraph 2 of its Statute, the Court held that a declaration

> must be interpreted as it stands, having regard to the words actually used [...] the Court cannot base itself on a purely grammatical interpretation of the text. It must seek the interpretation which is in harmony with a natural and reasonable way of reading the text [the words chosen in Iran's declaration were] a decisive confirmation of the intention of the Government of Iran at the time when it accepted the compulsory jurisdiction of the Court[35]

and that

> a declaration of acceptance of the compulsory jurisdiction of the Court, whether there are specified limits set to that acceptance or not, is a unilateral act of State sovereignty [...] The Court observes that the provisions of that Convention [VCLT] may only apply analogously to the extent compatible with the sui generis character of the unilateral acceptance of the Court's jurisdiction [...] declarations and reservations are to be read as a whole [...] since a declaration under Article 36, paragraph 2, of the Statute is a

[32] *Certain Norwegian Loans (France v. Norway), Judgment of 6 July 1957, I.C.J. Reports 1957, p. 9, 23*, also available at http://www.icj-cij.org/docket/files/29/4773.pdf (last visit 30 Jul 2015). Critical comments on the Court's interpretation with regard to the optional clause are noted, too. A Mark Weisburd, Failings of the International Court of Justice (OUP 2016) 112–117.

[33] *Military and Paramilitary Activities in and against Nicaragua (Nicaragua v. USA), Jurisdiction and Admissibility, Judgment of 26 November 1984, I.C.J. Reports (1984) p. 418, para. 60*, also available at http://www.icj-cij.org/docket/files/70/6485.pdf (last visit 30 Jul 2015).

[34] *Land and Maritime Boundary between Cameroon and Nigeria (Cameroon v. Nigeria: Equatorial Guinea intervening), Preliminary Objections, Judgment of 11 June 1998, I.C.J. Reports (1998) p. 291, para. 25*, also available at http://www.icj-cij.org/docket/files/94/7473.pdf (last visit 30 Jul 2015).

[35] *Anglo-Iranian Oil Co. (U.K. v. Iran), Preliminary Objection, Judgment of 22 July 1952, I.C.J. Reports 1952, p. 105 and 107*, also available at http://www.icj-cij.org/docket/files/16/1997.pdf (last visit 30 Jul 2015).

unilaterally drafted instrument, the Court has not hesitated to place a certain emphasis on the intention of the depositing State.[36]

It has been suggested that space law disputes, whether originating from the *corpus* of international space treaties regime or not, cannot be brought today and as the law stands before the Court for adjudication. With the greatest of respect, no deepest misconception has been advanced, for this suggestion is, at best, unfounded.

Given the authoritative pronouncement of the Court that 'the United Nations Charter contains no specific provision of itself conferring compulsory jurisdiction on the Court',[37] a more intimate look on the precise content of declarations in force becomes imperative.[38] It is the author's belief that assessing the actual practice of each State in being prepared to entrust the settlement of its disputes to the Court remains the utmost priority. Not surprisingly, efficient machinery is available.[39]

Furthermore, the author firmly believes that the Court enjoys the authority and jurisdiction to hear and try a space law dispute. A total of 72 States have deposited their declarations, albeit with reservations accompanying them.[40] One must note that 32 out of 72 States have established space agencies with certain capabilities.[41] From a slightly different perspective, 20 out of 72 States enjoy the so-called launch capability.[42]

It is well settled that every reservation must be given effect as it stands.[43] The Court interprets the relevant words of a declaration, including a reservation contained therein, in a natural and reasonable way. It pays due regard to the intention of the State concerned at the time when it accepted the compulsory jurisdiction of the

---

[36] *Fisheries Jurisdiction (Spain v. Canada), Jurisdiction of the Court, Judgment of 4 December 1998, I.C.J. Reports 1998, p. 453, paras 46–49*, also available at http://www.icj-cij.org/docket/files/96/7533.pdf (last visit 30 Jul 2015). Re-affirmed in *Whaling in the Antarctic (Australia v. Japan: New Zealand intervening), Judgment of 31 March 2014, p. 244, para. 36*, available at http://www.icj-cij.org/docket/files/148/18136.pdf (last visit 30 Jul 2015). Also, P Couvreur, The International Court of Justice and the Effectiveness of International Law (Brill 2017) 63.

[37] *Aerial Incident of 10 August 1999 (Pakistan v. India), Jurisdiction of the Court, Judgment of 21 June 2000, I.C.J. Reports (2000), p. 32, para. 48*, also available at http://www.icj-cij.org/docket/files/119/8088.pdf (last visit 30 Jul 2015).

[38] Inter alia, G Törber, The Contractual Nature of the Optional Clause (Hart 2015).

[39] *Contra* K Böckstiegel, Equipping the Court to deal with Developing Areas of International Law: Space Law in C Peck & R Lee (Eds.) (note 17) 447, G Meishan-Goh, (note 26).

[40] For the full list, please visit the official site of the Court at http://www.icj-cij.org/jurisdiction/index.php?p1=5&p2=1&p3=3 (last visit 30 Jul 2015).

[41] Australia, Belgium, Bulgaria, Canada, Costa Rica, Denmark, Egypt, Finland, Germany, Greece, Hungary, India, Ireland, Italy, Japan, Lithuania, Luxembourg, Mexico, the Netherlands, New Zealand, Nigeria, Norway, Pakistan, Peru, Philippines, Poland, Portugal, Romania, Spain, Sweden, Switzerland and the UK.

[42] Austria, Belgium, Denmark, Finland, Germany, Greece, India, Ireland, Italy, Japan, Luxembourg, the Netherlands, Norway, Poland, Portugal, Romania, Spain, Sweden, Switzerland and the UK.

[43] *Certain Norwegian Loans (France v. Norway), Judgment of 6 July 1957, I.C.J. Reports 1957, p. 27* (note 32). Re-affirmed in *Fisheries Jurisdiction (Spain v. Canada), Jurisdiction of the Court, Judgment of 4 December 1998, I.C.J. Reports 1998, p. 432, para. 47* (note 36).

Court and in a manner compatible with the effect sought by the reserving State. Therefore, one must examine the intentions of a reserving State. This task is, by all means, an onerous one. However, these intentions may be deduced not only from the text of the relevant clause. The context in which the clause is to be read and an examination of evidence regarding the circumstances of its preparation and the purposes intended to be served are crucial. In case an existing declaration has been replaced by a new declaration which contains a reservation, the intentions of the State may also be ascertained by comparing the terms of the two instruments.[44]

It appears that out of those 72 declarations, no State has included a *ratione materiae* reservation,[45] therewith excluding disputes originating from a conventional space law instrument or even customary international space law. This conclusion was reached by following strictly the relevant test set by the Court, when interpreting a quite usual term, namely, 'disputes arising out of or concerning […]' say space disputes. This language brings out the broad and comprehensive character of the formula employed. The wording of this format for reservations, in the Court's own words,

> […] exclude[s] not only disputes whose immediate 'subject matter' is the measures in question and their enforcement, but also those 'concerning' such measures and, more generally those having their 'origin' in those measures ('arising out of), that is to say those disputes which, in the absence of such measures, would not have come into being.[46]

No such undertaking has been deposited yet, and consequently, the Court enjoys a prima facie basis of jurisdiction with respect to space law disputes as between these 72 States. Indeed the compulsory jurisdiction of the Court is based on the law in force at the time of the institution of the proceedings, and it is never possible to

[44] *Fisheries Jurisdiction (Spain v. Canada), Jurisdiction of the Court, Judgment of 4 December 1998, I.C.J. Reports 1998, p. 432, paras 49–50* (note 36). Re-affirmed in *Whaling in the Antarctic (Australia v. Japan: New Zealand intervening), Judgment of 31 March 2014, ibid.*, (note 36). In the former case, the Court ruled that 'reservations from the Court's jurisdiction may be made by States for a variety of reasons; sometimes precisely because they feel vulnerable about the legality of their position or policy. Nowhere in the Court's case-law has it been suggested that interpretation in accordance with the legality under international law of the matters exempted from the jurisdiction of the Court is a rule that governs the interpretation of such reservations […] The fact that a State may lack confidence as to the compatibility of its actions with international law does not operate as an exception to the principle of consent to the jurisdiction of the Court and the freedom to enter reservations' *(para. 54)*. The Court continued by upholding its finding in the *Military and Paramilitary Activities in and against Nicaragua (Nicaragua v. U.S.A.), Jurisdiction and Admissibility, Judgment of 26 November 1984, p. 418, para. 59* (note 33) and repeated that 'declarations of acceptance of the compulsory jurisdiction of the Court are facultative, unilateral engagements, that States are absolutely free to make or not to make. In making the declaration a State is equally free either to do so unconditionally and without limit of time for its duration, or to qualify it with conditions or reservations'.

[45] Inter alia, S Alexandrov, Reservations in Unilateral Declarations Accepting the Compulsory Jurisdiction of the International Court of Justice (Martinus Nijhoff 1995). Conditions and other 'generic' types of reservations (e.g. of domestic jurisdiction, temporal, multilateral, commonwealth, war, etc.) remain outside the ambit of this presentation.

[46] *Fisheries Jurisdiction (Spain v. Canada), Jurisdiction of the Court, Judgment of 4 December 1998, I.C.J. Reports 1998, p. 432, para. 62* (note 36).

determine *in abstracto* whether a case of compulsory jurisdiction exists, without studying the application pending.[47] But this is exactly the task an able practitioner of international law is commissioned to undertake.

9.5 Is There a Need for a World 'Space' Court?

Some academics hold the view that a sector-specific dispute settlement mechanism is urgently needed. They argue that space activities take place under unique circumstances. The proliferation of space activities, they continue, is essentially connected with utterly specialized knowledge, and international community would be benefited from the tension reduced between various actors. They also advance a dogmatic argument, namely, the ability to respond more timely and efficiently to changes in the field, ergo allowing further elaboration of space law. Lastly, they invoke the enforcement of the rule of law in outer space as the most important reason for the establishment of a 'sectorialized dispute settlement mechanism'. For all these reasons, they conclude, the general jurisdiction entrusted by the international community to the Court is rather unsuitable for the settlement of space-related disputes.[48]

Convincing or not, these arguments merit our full attention. After all, two successive Presidents of the Court – Stephen Schwebel and Gilbert Guillaume – stated their appreciations against the proliferation of international tribunals.[49] Eminent scholars have taken a similar stand, too.[50] In addition, the 2006 Report of the ILC Study Group on Fragmentation of International Law indicates that it is undesirable to establish a general space court.[51]

Some comments need to be registered. In the authors' view, there is a certain degree of uniqueness in each and every case, whether brought before the Court for adjudication or not. Since the Court does not operate in complete isolation, a

---

[47] S Rosenne, (note 31) 794.

[48] G Meishan-Goh (note 26) 139, 169–191, I Diederich-Verschoor, 'The Settlement of disputes in Space: New Developments', 26 J. Sp. L. (1998) 41–9. Also, ILA: Report of the Space Law Committee (Helsinki 1996 & Taipei 1998), M Williams (Rapporteur) 457–476 & 239–272, respectively.

[49] In short, their main arguments were cases of overlapping jurisdictions, forum shopping, serious risk of conflicting jurisprudence and fragmentation of international law. Brown believes that limiting factors on the development of cross-fertilization of rules are traced in the particular drafting of constituent instruments and the fact that international courts may have their own specific agendas and functions, C Brown, A Common Law of International Adjudication (OUP 2009) 235.

[50] For an overview, inter alia, H Caminos, The Growth of Specialized International Tribunals and the Fears of Fragmentation in International Law in N Boschiero, T Scovazzi et al. (Eds.), International Courts and the Development of International Law (Springer 2013) 55, where the views expressed by Judge Shigeru Oda, Judge Rosalyn Higgins, Jonathan Charney, Shabtai Rosenne, Sir Robert Jennings and Bruno Simma. A Lang, 'The Role of the International Court of Justice in a Context of Fragmentation' (2013) 62 ICLQ 777.

[51] 2006 YBILC II, Part 2.

cautious[52] approach is truly needed. Moreover, as an authority on the Court's practice suggested some time ago, crisis management is, at most, an auxiliary role for the Judiciary.[53]

During the last 25 years, the record is abundantly clear: the Court has not abstracted itself out of the *exelixis* of the community it serves. On a number of disputes, it has juxtaposed the traditional structure of international law (bilateral par excellence) with the notion of 'common', 'collective' or 'general interest'.[54] No other entity but the Court has shown its unparalleled resolution to draw on the work of specialized institutions: after reviewing their projects and conclusions, it uses their expertise to bolster its own adjudicatory role. With regard to evidence, as Judge Van Eysinga explained in his *Separate Opinion* in the *Oscar Chinn case*, 'the Court is not tied to any system of taking evidence [...] its task is to cooperate in the objective ascertainment of the truth'.[55]

The Court seems to have never refused to hear 'experts', on a contentious case.[56] The experts' assessments were always taken under serious consideration. As for the time element, given that justice delayed is justice denied, one has only to recall the practice of the Court, especially when provisional measures are asked to be indicated.

This brings the discussion to the substance of the matter: development of space law rules and principles does not derive only from contentious cases. What about the instructive role of advisory proceedings?

Right from the outset, an advisory opinion asked in the proper fashion by the UN General Assembly or the Security Council may be of extreme value. If the Court follows its jurisprudence, namely, that it will not tolerate the abuse of the advisory procedure and shall apply the so-called '*Eastern Carelia*' principle,[57] issues of space law of wider concern, actively engaging the work of the requesting UN organ or the interest of the international community, may be successfully tackled. What Judge Gros maintained in his *Declaration* in the *Western Sahara Advisory Opinion* remains valuable, for, 'when the Court gives an advisory

---

[52] Judge P Kooijmans, 'The ICJ in the 21st Century: Judicial Restraint, Judicial Activism, or Proactive Judicial Policy' (2007) 56 ICLQ 741.

[53] S Rosenne, A Role for the International Court of Justice in Crisis Management? (note 11) 61.

[54] V Gowlland-Debbas, Judicial Insights into Fundamental Values and Interests of the International Community in A. Müller, D Raič & J Thuranszky (Eds.) (note 1) 349.

[55] *Oscar Chinn Case, PCIJ Ser. A/B No 63 (1934) 146*, also available at http://www.icj-cij.org/pcij/serie_AB/AB_63/05_Oscar_Chinn_Opinion_Eysinga.pdf (last visit 30 Jul 2015).

[56] Articles 50–51 of the Statute of the Court and articles 62 & 67 of its Rules of the Court are quite clear. An excellent example can be found in the *Whaling in the Antarctic (Australia v. Japan: New Zealand intervening) Judgment of 31 March 2014, para. 20–22* (note 36). A Riddel & B Plant, Evidence before the International Court of Justice (B.I.I.C.L. 2009) 62. M Mbengue, 'Scientific Fact-finding by International Courts and Tribunals' (2012) 3 JIDS 509. C Foster, 'New Clothes for the Emperor? Consultation of Experts by the International Court of Justice' (2014) 5 JIDS 139.

[57] *Status of Eastern Carelia, Advisory Opinion of 23 July 1923, PCIJ Ser. B No 5, 28*, also available at http://www.icj-cij.org/pcij/serie_B/B_05/Statut_de_la_Carelie_orientale_Avis_consultatif.pdf (last visit 30 Jul 2015).

opinion on a question of law it states the law [...] the absence of binding force does not transform the judicial operation into a legal consultation, which may be made use of or not according to choice'.[58]

After all, there is no doubt the Court is under a prima facie obligation to reply[59] and enjoys wide discretion as per ascertaining[60] and reformulating[61] questions transmitted to it. It is also imperative for the Court to have a clear picture of the specific situation.[62] Would that be the case, experience of the Court's practice manifests its ability to react promptly – not at the expense of quality of reasoning. The Court remains at the service of international community in its entirety and has been 'radiating through the entire global community a consciousness of the international rule of law'.[63]

Lastly, and no less importantly, no State shares the view of establishing a World Space Court, for – to the best of the author's knowledge – no such proposal has been ever advanced or recorded.

9.6 Concluding Remarks

So allow me to conclude with some more constructive observations. Although they may appear a little eccentric, they are not views for which the author is able to claim originality. Since '[g]overnments do not want philosophers as their legal advisers [...] I have no particular theory to propound to you',[64] and '[...] even the counsel

---

[58] *Western Sahara, Advisory Opinion of 16 October 1975, Declaration of Judge Gros, I.C.J. Reports (1975), p. 73, para. 6*, also available at http://www.icj-cij.org/docket/files/61/6199.pdf (last visit 30 Jul 2015).

[59] *Interpretation of Peace Treaties with Bulgaria, Hungary and Romania (First Phase), Advisory Opinion of 30 March 1950, I.C.J. Reports (1950), p. 71*, also available at http://www.icj-cij.org/docket/files/8/1863.pdf (last visit 30 Jul 2015). Followed by *Judgment No 2867 of the Administrative Tribunal of the International Labour Organization upon a Complaint filed against the International Fund for Agricultural Development, Advisory Opinion of 1 February 2012, I.C.J. Reports (2012), p. 48, para. 96*, also available at http://www.icj-cij.org/docket/files/146/16871.pdf (last visit 30 Jul 2015).

[60] *Interpretation of the Agreement of 25 March 1951 between the WHO and Egypt, Advisory Opinion of 20 December 1980, I.C.J. Reports (1980), p. 88, para. 35*, also available at http://www.icj-cij.org/docket/files/65/6303.pdf (last visit 30 Jul 2015).

[61] *Legal Consequences of the Construction of a Wall in the Occupied Palestinian Territory, Advisory Opinion of 9 July 2004, I.C.J. Reports (2004), p. 156, para. 44*, also available at http://www.icj-cij.org/docket/files/131/1671.pdf (last visit 30 Jul 2015).

[62] C Greenwood, Judicial Integrity and the Advisory Jurisdiction of the International Court of Justice in G Gaja & J Stoutenburg (Eds), Enhancing the Rule of Law through the International Court of Justice, (Brill/Nijhoff 2012) 68.

[63] C Weeramantry, The International Court of Justice at 50 – Introduction in C. Peck & R. Lee (Eds.) (note 17) 1.

[64] S Rosenne, The Perplexities of Modern International Law (Martinus Nijhoff 2004) 2. Also, S Bouwhuis, 'The Role of an International Legal Adviser to Government' (2012) 61 ICLQ 939.

whom states hire can be quite pragmatic, faced with the imperative of winning cases rather than losing bravely but with the academic laurels intact'.[65]

The Court does not sit at the top of any jurisdictional pyramid. It does not enjoy the prerogative to adjudicate at will a preferential set of disputes. It is established to try *any* dispute whatsoever and produce a fair and objective reasoned solution to it. Its job is to apply the law. This may explain why it faced many novel situations and rendered, on a number of disputes, bold decisions while maintaining a careful balance between competing considerations. It has also assisted States and the international community as a whole to change international law, without derogating from its function and role. In addition and as it was stressed by Sir Hersch Lauterpacht, '[i]nstitutions set up for the achievement of definite purposes grow to fulfil tasks not wholly identical with those which were in the minds of their authors at the time of their creation'.[66]

The Court seized many times the opportunity to shape and polish the law. It is an essential part of the process, and there is widespread agreement on this. In addition, the lawyers appearing before the Court bring a common legal experience. They also contribute to the proper administration of international justice.

Access to the Court has become 'universal in nature'.[67] Space law is a new field indeed; settlement of disputes is most certainly not.

Although there is no concrete and hard evidence to support the view that a multiplicity of international tribunals has impaired the unity of jurisprudence, what the legal audience focuses on is the influence of the Court on the system of substantive law. As a general matter, its pronouncements are presumed to be of great value, notwithstanding its alleged role of the 'gatekeeper', that is to say overseeing and controlling the evolution of certain rules and principles of international law. The Court does not exercise a monopoly over the guardianship of the rule of law but shares a concurrent responsibility with all other actors in the field.[68]

International instruments regulate a variety of space features. Devising and revising the necessary structures, with respect to space activities, require global commitment, close co-operation and amicable contribution between all organs of the UN system and, most prominently, between States themselves. Future cases dealing with issues of space law should be referred to the World Court as a full Court, for a wide and comprehensive experience will be needed.

In April 2016 the International Court of Justice celebrated 70 years of activity. The event is of undoubted significance as no other international judicial body has

---

[65] Sir Franklin Berman, (note 29) 13.

[66] Sir Herch Lauterpacht, The Development of International Law by the International Court (CUP 2010) 5.

[67] A Zimmermann, Between the Quest for Universality and its Limited Jurisdiction: The Role of the International Court of Justice in Enhancing the International Rule of Law in G Gaja & J Stoutenburg (Eds), (note 62) 34.

[68] M Shahabudeen, The World Court at the turn of the Century in A. Müller, D Raič & J Thuranszky (Eds.) (note 1) 8. Also, C Romano, 'A Taxonomy of International Rule of Law Institutions' (2011) 2 JIDS 241.

been so long-lived. Still, no space-related case has been recorded in its docket. Should any such case arise, States will appear before it. Such conduct will justify both loyalty and confidence in it. That much States owe to the Court.

As Nicolas Politis observed in 1920:

> [t]he obligation of recourse to a Court is above all a question of confidence; and confidence either does or does not exist.[69]

[69] League of Nations, Document Concerning the Action Taken by the Council of the League of Nations Under Article 14 of the Covenant and the Adoption by the Assembly of the Statute of the Permanent Court (1921) 243.

Chapter 10
The Legacy of the Dinosaurs: Regulation of Planetary Defence and Near-Earth Objects at a Global Level

George Leloudas, Michael Chatzipanagiotis, and Konstantina Liperi

In recent years, Near-Earth Objects (NEOs) have caused increasing concerns among space scientists and policymakers. While attacks on Earth from NEOs are rare, the atmospheric impact of an about 20-m asteroid near Chelyabinsk, Russia, in February 2013 served as a warning that asteroid/comet strikes could seriously threaten humanity.[1] Planetary defence requires strong international co-operation. Despite the initiatives at UN COPUOS, international law falls considerably short of addressing issues pertaining to the detection and mitigation of hazardous NEOs.

This paper builds upon the "risk society" of Ulrich Beck to propose the creation of a Convention on Planetary Defence. The convention will address co-operation issues among States and, moreover, establish a Global Fund for Planetary Defence. This Fund will be used for R&D on NEOs, for identifying best practices in deterring potential attacks, and for compensating relevant damages.

It is suggested that the convention creates a stand-alone international organisation that operates the Fund to fulfil the aforementioned objectives. Parallels will be drawn from various existing international/national arrangements dealing with disasters, such as terrorism and nuclear and pollution accidents. Yet, it will be demonstrated

[1] Secure World Foundation, *Near Earth Objects: Responding to the International Challenge* (April 2014) p. 2, online: http://swfound.org/media/170684/SWF_NEOs-Responding_to_the_International_Challenge_2014.pdf (last accessed on 15 January 2018).

G. Leloudas
Institute of International Shipping and Trade Law, Swansea University, Swansea, UK
e-mail: g.leloudas@swansea.ac.uk

M. Chatzipanagiotis (✉)
University of Cyprus, Nicosia, Cyprus
e-mail: mchatz07@ucy.ac.cy

K. Liperi
Department of Civil Aviation, Ministry of Transport, Communications and Works, Nicosia, Cyprus
e-mail: kliperi@mcw.gov.cy

G. D. Kyriakopoulos, M. Manoli (eds.), *The Space Treaties at Crossroads*,
https://doi.org/10.1007/978-3-030-01479-7_10

that the unique risk of NEOs requires a sui generis solution. Furthermore, liability and recourse issues, including governmental liability for failing to mitigate NEO strikes, will be analysed.

10.1 The Threat of NEOs

10.1.1 NEOs and PHOs

Near-Earth Objects (NEOs) are planetary debris, created from the formation of our solar system some 4.6 billion years ago. They are mostly asteroids and comets that have been nudged by the gravitational attraction of nearby planets into orbits that intersect that of the Earth or are within 0.3 astronomical units.[2]

These celestial bodies are the remnants of the primitive building blocks of the planets, called "planetesimals".[3] In essence, our giant planets were formed by the agglomeration of the planetesimals through several collisions. The asteroids and comets are the leftover pieces that remained in orbit around the Sun.

Comets are composed mostly of water ice with embedded dust particles and were originally formed in the cold outer planetary system. Asteroids are rocky or metallic bodies formed in the warmer inner solar system between the orbits of Mars and Jupiter. The composition, density, and size of NEOs vary greatly and influence their behaviour when entering the Earth's atmosphere and their reaction to countermeasures.[4]

A potential hazardous object is an object in our solar system that passes within 0.05 AU (about 7.5 million km) of Earth's orbit and is large enough, i.e. at least 50 m in diameter to pass through Earth's atmosphere and cause significant damage on impact.[5]

---

[2] Astronomers call the Earth to Sun distance one "astronomical unit". See on definition of NEOs *UK Task Force on potentially hazardous Near Earth objects*, Report (September 2000), pp. 11–12, online: NSS http://www.nss.org/resources/library/planetarydefense/2000-ReportOfTheTaskForceOnPotentiallyHazardousNearEarthObjects-UK.pdf (last accessed on 15 January 2018).

[3] International Academy of Astronautics, *Dealing with the threat to earth from asteroids and comets*, Edited by I. Bekey, p.16, online: http://swfound.org/media/167468/Dealing_With_Threat_To_Earth_Asteroids_Comets.pdf (last accessed on 15 January 2018).

[4] Ibid.

[5] NASA, *Near-Earth Object Survey and Deflection Analysis of Alternatives*, Report to the Congress (March 2007), p. 7, online: http://www.nasa.gov/pdf/171331main_NEO_report_march07.pdf (last accessed on 15 January 2018).

10.1.2 The Disastrous Potential

The odds of a cosmic collision happening during a human life span are significantly limited. Threats arising from potential diseases, natural disasters, or auto accidents are much greater rather than a threat from a NEO impact.[6] In fact, the vast majority of people are oblivious to the prospect of an asteroid or a comet strike.

However, what are very tiny risks for impacts during a human lifetime become certainties on geologic timescales. The increasing understanding of the magnitudes and qualitative features of environmental consequences of impacts of objects suggests that prime attributes of impacts, not duplicated by any other natural processes, are (1) extreme suddenness, providing little opportunity for escape and no chance for adaptation, (2) globally pervasive, and (3) unlimited potential for overwhelming destruction of the life-sustaining characteristics of the fragile ecosphere.[7] Our planet's geological and biological history reveals that a series of destructive impact events have occurred throughout the years. Therefore, in recent years, the issue of planetary defence has become an increasing area of concern among the world's space scientists, astronomers, and policymakers.

The now widely accepted theory that a NEO strike is responsible for the ultimate destruction of non-avian dinosaurs 65 million years ago serves as a strong illustration of the disastrous potential such an impact may have. Yet, there are also relatively recent impact events. In 1908 an object, believed to be an asteroid or a comet, exploded over Tunguska River in Russia causing the destruction of 2000 km^2 of Siberian forest. The overhead detonation was equivalent to the explosive force of a 4 megaton bomb.[8] In 2013, a 17–20-m-sized orbiting asteroid entered the atmosphere over Chelyabinsk, Russia, and exploded with a force of 500 kilotons. The impact event caused injuries to about 1600 people, with some of them having to be hospitalised from cuts caused by flying materials.[9]

If the Tunguska or Chelyabinsk event had happened over a densely populated area of the world, it would have a catastrophic effect causing the death of millions of people. The fact that these two events happened within the relatively short period of 100 years must serve as a wake-up call for the international community to adopt collective and holistic measures.

---

[6] Center for NEO Studies (CNEOS), online: https://cneos.jpl.nasa.gov/about/target_earth.html (last accessed on 15 January 2018).

[7] See details in C. Chapman, *Impact lethality and risks in today's world: Lessons for interpreting Earth history* in C. Koeberl and K. MacLeod. (eds), *Catastrophic Events and Mass Extinctions: Impacts and Beyond*, Geological Society of America Special Paper 356, Colorado 2002, pp. 7–19.

[8] Secure World Foundation, *Near Earth Objects: Responding to the International Challenge*, April 2014 p. 6, online: http://swfound.org/media/170684/SWF_NEOs-Responding_to_the_International_Challenge_2014.pdf (last accessed on 15 January 2018).

[9] Ibid.

10.2 Methodology of Threat Mitigation

Unlike the dinosaurs' age, recent developments in space science and technology enable humanity to take preventive action. Scientists can predict whether an object can actually serve as a threat by detecting and categorising NEOs. Technology also allows for the successful prevention or mitigation of a hazardous NEO's impact.

10.2.1 Detection of NEOs

The detection and tracking of NEOs is the first and fundamental step in designing efficient mitigation strategies. To "know your enemy" enables appropriate planning for its defeat. The orbit of an object defines if, when, and how a strike may occur, as well as the warning times and deflection requirements.[10]

The identification of NEOs can be done with the aid of ground-based or space-based systems.

Ground-based systems use large field of optical sensors to scan for NEOs during the night. They cannot function during daylight or twilight, and their operation can be interfered by weather, atmospheric turbulence, scattering from moonlight, and atmospheric attenuation. Nevertheless, ground systems, compared to space-based ones, are easier to build, verify, operate, maintain, and upgrade.[11] NEO identification is currently being achieved by a range of ground instruments. The development of Space Surveillance Telescope[12] is anticipated to bridge gaps in observational coverage by enabling the detection and tracking of small objects in deep space.[13]

Space-based systems are capable to scan the entire sky at any given time without any interference.[14] It should be noted, however, that space-based systems are more expensive to develop, and they run the risk of launch and deployment failure. Additionally, space-based systems are likely to have shorter lifetime. This may

---

[10] Small Bodies Assessment Group (SBAG), *Goals Document* Goal II: Planetary Defense, Draft version 0.5 (26/6/2015), p. 1, online: http://www.lpi.usra.edu/sbag/goals/Goal_II_sbag_pdg_draft0.5.pdf (last accessed on 15 January 2018).

[11] NASA, *supra* note 5, p. 8.

[12] It is developed by the Massachusetts Institute of Technology's Lincoln Laboratory for the Defense Advanced Research Projects agency and the US Air Force. For more info, see http://www.darpa.mil/program/space-surveillance-telescope (last accessed on 15 January 2018).

[13] Chile is also developing the Large Synoptic Survey Telescope which is anticipated to be operational in 2020. More information herein online: Large Synoptic Survey Telescope http://www.lsst.org/ (last accessed on 15 January 2018).

[14] Some of the current space-based instruments are NASA's NEOWISE and Canada's NEOSSAT; see infra para. 3.2.

compromise the objectives of a NEO threat mitigation strategy, which may require the tracking of NEOs for extended periods of time.[15]

After identifying the objects, their physical characteristics are determined. This is vital in order to predict the impact risk on Earth, as well as to calculate the object's response to a mitigation mission. The characterisation of an object involves the determination of its composition, size, shape, and rotation rate.[16] Radar observation is a powerful tool for characterising NEOs.

10.2.2 Prevention or Mitigation of the Impact

The most effective method to avert a NEO impact on Earth is to change its orbit. This can be achieved by using energy to change its velocity, so that it arrives sooner or later than Earth at the point of intersection.[17] With advance warning of many years, a change in the velocity of a NEO by only few centimetres per second can successfully cause the object to miss the Earth.[18] Several techniques have been proposed over the years to shift a NEOs' trajectory.

One such technique is the *kinetic impactor*, according to which a NEO is deflected after an impact by a spacecraft, which causes a slight shift of the orbit. In essence, the spacecraft hits the NEO at high relative velocity, so that the object's speed immediately changes because of the energy and momentum exchange.[19] Another technique is the *gravity tractor*, in which a spacecraft flies alongside a NEO for a certain period of time, gradually pushing it off course through the small gravitational attraction between the two. Moreover, the *blast deflection* technique requires the use of nuclear explosives near the NEO causing its outer layers to evaporate, changing in that way its trajectory. Non-nuclear explosives can also be used to perform small or medium pushes to the NEO.[20]

For the successful prevention or mitigation of a menacing NEO impact, time is of essence. In order to be able to plan and deploy a successful deflection mission, three elements are necessary: (a) sufficient advance warning of a minimum 10–15 years, (b) deflection capability, and (c) an international decision-making mechanism.[21] These three elements enable the deflection campaign to proceed timely, so that the predicted impact is prevented.

---

[15] NASA, *supra* note 5, p. 9.

[16] National Aeronautics and Space Administration (NASA), Planetary defence, online: https://www.nasa.gov/planetarydefense/faq.

[17] International Academy of Astronautics, *supra* note 3, p. 7.

[18] *Ibid.*, p. 8.

[19] International Academy of Astronautics, *supra* note 3, p. 8.

[20] Secure World Foundation, *supra* note 8, pp. 13–14.

[21] Association of Space Explorers, *Asteroid Threats: A Call for Global Response* (25 September 2008), p. 14, online: Space Explorers http://www.space-explorers.org/ATACGR.pdf (last accessed on 15 January 2018).

When preparing to prevent an impact, the communication with the public is also of utmost importance. A coordination plan should be developed detailing the nature of the threat, evacuation activities, and instructions to the public. The aftermath of the Indian Ocean Tsunami and Hurricane Katrina revealed that only realistic planning and preparation can minimise the chaos that is likely to occur.[22]

10.3 Current Efforts for Threat Mitigation

Currently, efforts for NEO threat mitigation are undertaken mainly by international organisations and individual States. At the same time, private initiatives have been developed.

10.3.1 International Organisations

10.3.1.1 United Nations

The United Nations (UN) have been undertaking significant work on NEOs, following the report of the Third United Nations Conference on the Exploration and Peaceful Uses of Outer Space (UNISPACE III) in 1999,[23] which encouraged the improvement of international co-operation on this issue. As a follow-up to this report, the UN Committee on the Peaceful Uses of Outer Space (COPUOS) created the Action Team on Near-Earth Objects, also referred to as Action Team 14 (AT-14),[24] which comprises scientific and space agency delegates from interested States and non-governmental organisations.

In 2013, the UN General Assembly approved a series of concrete measures,[25] to protect the Earth against hazardous NEOs. This important milestone resulted from a series of lengthy discussions in the UN COPUOS and AT-14. The measures approved involve, inter alia, the creation of an International Asteroid Warning Network (IAWN) and a Space Mission Planning Advisory Group (SMPAG).[26]

---

[22] International Academy of Astronautics, *supra* note 3, p. 9.

[23] Report of the Third United Nations Conference on the Exploration and Peaceful Uses of Outer Space, Vienna, 19–30 July 1999 (18 October 1999), A/CONF.184/6.

[24] Report of the Committee on the Peaceful Uses of Outer Space, Fifty-sixth Session (2001), A/56/20.

[25] Resolution adopted by the General Assembly on 11 December 2013 (16 December 2013), A/RES/68/7520.

[26] These measures were initially proposed by the Association of Space Explorers (ASE) in their report, *Asteroid Threats: A Call for Global Response* (2008); see *supra* note 21. ASE recommended that the UN should support asteroid warning and information sharing, mission operations planning, as well as authorisation and oversight functions. The report was submitted to UN

International Asteroid Warning Network

The IAWN is tasked with the coordination of an international group of organisations involved in detecting, tracking, and characterising NEOs. IAWN also acts as an international coordination service for processing alerts about possible threats. Furthermore, IAWN will develop a strategy to assist governments in the analysis of asteroid impact consequences and the planning of mitigation responses. Currently, there are eight official signatories to the IAWN Statement of Intent.[27]

Space Mission Planning Advisory Group

The SMPAG is comprised of voluntary representatives of Member States with space agencies. Its purpose is to prepare an international response to a NEO impact threat by exchanging information, promoting collaborative research and mission opportunities, as well as planning activities for NEO threat mitigation.[28] Since 2016, as per the General Assembly resolution 71/90, the United Nations Office for Outer Space Affairs serves as the permanent secretariat to SMPAG.[29]

10.3.1.2 European Union

In the framework of the European Union, the European Commission has funded the NEOshield project which aims at improving our knowledge on NEOs' physical characteristics and investigating the effectiveness of asteroid threat-reduction techniques. The NEOshield project was undertaken by research institutes, universities, and industrial partners in Germany, France, the UK, Spain, and in the USA and Russia. Since 2015, the work continues with NEOshield-2 which aims to investigate in more detail technologies critical to NEOs' deflection missions as well as to refine NEOs' characterisation.[30]

COPUOS in 2009 for consideration and subsequent action. The UN COPUOS Scientific and Technical Subcommittee, its NEO Working Group, and AT-14 discussed and refined these proposals.

[27] More information herein online: United Nations Office for Outer Space Affairs http://www.unoosa.org/oosa/en/ourwork/topics/neos/iawn.html (last visited 15 January 2018).

[28] See SMPAG Terms of Reference, online: http://www.cosmos.esa.int/web/smpag/terms-of-reference-v0 (last accessed on 15 January 2018).

[29] United Nations Office for Outer Space Affairs, *supra* note 27.

[30] See, for instance, the NEOshield-2 project. More information herein online: http://www.neoshield.eu/science-technology-asteroid-impact/ (last accessed on 15 January 2018).

10.3.1.3 European Space Agency

The European Space Agency (ESA) has developed a multisegment Space Situational Awareness (SSA) programme. Its aim is to detect, predict, and assess the risk to life and property due to man-made space debris objects, harmful space weather, and potential impacts of NEOs. The objective of SSA's Near-Earth Object segment is to raise awareness of the current and future position of NEOs in relation to Earth, to estimate the likelihood of impacts, to assess the consequences of any possible impact, and to develop NEO deflection methods.[31] The data collected from telescopes is retrieved by the NEO Coordination Centre (NEOCC). In the event of high-risk impact predictions, the data will be cross-checked with NASA's SENTRY system which is operated by the Jet Propulsion Laboratory (JPL).[32]

Moreover in 2014, Philae, a robotic lander, successfully achieved the first ever landing on the surface of a comet, as part of ESA's Rosetta mission. Until the end of the mission in 2016, Philae took images which will assist scientists to study and learn more on the composition of comets and on their role in the evolution of the Solar System. This knowledge will significantly help in the planning of threat mitigation techniques.

10.3.2 Individual States

10.3.2.1 United States of America

The US National Aeronautics and Space Administration (NASA) maintains a leading role in the prevention of NEOs' impact. Planetary defence has been gaining particular attention in the USA, and it is within the nation's top-tier space policy issues.[33]

Already in 1998, the US Congress directed NASA to locate at least 90% of all NEOs with a diameter of 1 km or greater within the period of 10 years. NASA and its partners managed to achieve this goal in 2011. Additionally, NASA's Authorization Act of 2005[34] set a far more challenging task for NASA, i.e. to detect, track, catalogue, and characterise 90% of all NEOs with a diameter of 140 m or greater by 2020.[35]

---

[31] European Space Agency (ESA), more information herein online: http://www.esa.int/Our_Activities/Operations/Space_Situational_Awareness/Near-Earth_Objects_-_NEO_Segment (last accessed on 15 January 2018).

[32] Ibid.

[33] See National Space Policy of the United States of America (28 June 2010) p. 4, online: http://history.nasa.gov/national_space_policy_6-28-10.pdf (last accessed on 15 January 2018) and NASA Strategic Plan 2014, p.8, online: http://www.nasa.gov/sites/default/files/files/FY2014_NASA_SP_508c.pdf (last accessed on 15 January 2018).

[34] Also known as the George E. Brown, Jr. Near-Earth Object Survey Act.

[35] Section 321 of the NASA Authorization Act of 2005 (Public Law No. 109–155). More information herein online: http://www.gpo.gov/fdsys/pkg/PLAW-109publ155/pdf/PLAW-109publ155.

The vast majority of NEO discoveries have been the result of NASA-supported ground-based telescopic surveys including the Catalina Sky Survey (CSS) and Spacewatch near Tucson, Arizona; the LINEAR project near Socorro, New Mexico; Pan-STARRS1 on Haleakala, Maui, Hawaii; LONEOS near Flagstaff, Arizona; and the NEAT project run by NASA/JPL. Also the NEOWISE project has been reactivated to discover and characterise asteroids using a near-infrared space telescope in an Earth polar orbit.[36] The technological progress and the appropriate funding have enabled increased discovery of NEOs over the last years. This is demonstrated by the large NEO catalogue maintained by the Minor Planet Center (MPC) in Cambridge, Massachusetts, which collects and distributes data to the international community.[37]

In 2016, NASA established a Planetary Defense Coordination Office (PDCO), with the mission to safeguard the early detection of potentially hazardous objects – asteroids and comets with a size capable to reach Earth's surface and whose orbits can bring them within 0.05 astronomical units of Earth. The PDCO uses data from projects supported by NASA's Near-Earth Object (NEO) Observations Program.[38]

10.3.2.2 Canada

In 2013, Canada launched the Near-Earth Object Surveillance Satellite (NEOSSat). It is the world's first space telescope dedicated to detecting and tracking asteroids and satellites. It performs a circle of the globe every 100 min, scanning space near the Sun for hazardous asteroids. Thanks to its location, it can spot asteroids which are difficult to track with ground telescopes, and it can operate 24/7 without any limitations.[39]

10.3.2.3 Russian Federation

The Russian Federation is working on an SSA programme to detect, inter alia, hazardous NEOs. The International Scientific Optical Network (ISON), a growing international network of small telescopes linked together, is a major contributor in establishing an SSA programme. Additionally, Russia also operates several radar telescopes for characterisation of identified asteroids while planning to build

---

pdf (last accessed on 15 January 2018). The programme, however, has been unsuccessful so far, owing to administrative and budget restraints. More information herein online: http://phys.org/news/2014-09-nasa-asteroid-defense-falls-short.html (last accessed on 15 January 2018).

[36] See CNEOS, NEO Search Programme. More information herein online: http://neo.jpl.nasa.gov/programs/intro.html (last accessed on 15 January 2018).

[37] Secure World Foundation, *supra* note 8, p. 8.

[38] More information herein online: NASA, Planetary Defence Coordination Office: https://www.nasa.gov/planetarydefense/overview (last accessed on 15 January 2018).

[39] More information herein online: Canadian Space Agency, online: http://www.asc-csa.gc.ca/eng/satellites/neossat/ (last accessed on 15 January 2018).

several new ones. Russia is also looking to build and launch a space-based NEO telescope by 2021.[40]

10.3.2.4 Japan

Japan has successfully completed the Hayabusa mission, which collected samples from a small asteroid named Itokawa and returned them to Earth in June 2010. In 2014 the Japan Aerospace Exploration Agency (JAXA) successfully launched its Hayabusa2 mission to rendezvous with an asteroid, land a small probe plus three mini rovers on its surface, and then return samples to Earth. NASA and JAXA are cooperating on the science of the mission.[41]

10.3.3 Private Initiatives

The Spaceguard Foundation is a private, non-profit international organisation, set up in 1996 and based in Italy. Its objective is to study, discover, and observe NEOs and to protect the Earth environment against NEO impact threats.[42]

The *Emergency Asteroid Defence Project* (EADP) is a Danish non-governmental organisation, which includes scientists, engineers, and other interested persons trying to raise $200k on crowdfunding website Indiegogo, to continue its research into hypervelocity asteroid intercept vehicles (HAIVs) designed to deflect or disperse asteroids and comets with only a few days' warning.[43]

10.3.4 Evaluation of Current Efforts

All these efforts move to the correct direction and represent significant steps for alleviating the threat posed by NEOs. Nevertheless, they are inefficient, because they lack coordination, appropriate funding, and technical resources. In this regard, one could mention the 2014 Report in NASA's Inspector General[44] which states that

[40] Secure World Foundation, *supra* note 8, p. 8–9.

[41] See NASA Science Beta, Japan Launches Asteroid Mission, (4 December 2014) online: http://science.nasa.gov/science-news/science-at-nasa/2014/04dec_hayabusa2/ (last accessed on 15 January 2018).

[42] See The Spaceguard Foundation, online: http://spaceguard.rm.iasf.cnr.it/SGF/INDEX.html (last accessed on 15 January 2018).

[43] See Emergency Asteroid Defence Project, online: http://eadproject.com/ (last accessed on 15 January 2018).

[44] NASA's Efforts to Identify Near-Earth Objects and Mitigate Hazards, Audit report (15 September 2014) online: https://oig.nasa.gov/audits/reports/FY14/IG-14-030.pdf (last accessed on 15 January 2018).

due to administrative and budget restraints as well as organisational weaknesses, NASA will not be able to detect, track, catalogue, and characterise 90% of all NEOs with a diameter of 140 m or greater by 2020 as required by the NASA's Authorization Act of 2005.

Planetary defence is a global and complicated issue and at the same time crucial for the well-being and the survival of our species. Therefore, efforts should be organised at a global scale.

10.4 Proposal for an International Approach

10.4.1 Background: Risk Perception and Risk Management

Eric Posner, writing in 2004, argued that the need for organising a planetary defence system is underestimated despite the catastrophic potential of NEOs.[45] The reasons behind this trend are "... a combination of the science-fiction label that has become attached to the menace, the lack of any history of casualties being caused by asteroids..., and the sense that concern with such low-probability dangers would be a distraction from the struggle against international terrorism".[46]

His findings should not come as a surprise to the scientific community. A survey, which took place in 1992 in the USA, revealed "modest support for detection efforts but considerable opposition to the use of weapons in space, even to deflect a threatening asteroid. The survey respondents indicated a strong preference for collecting more data on the risk before developing a defence system".[47] This preference run (partially) against the perception among the same respondents that NEO risks were "extreme with regard to being unknown to scientists and the public, distant in time (non-immediate), uncontrollable and catastrophic".[48]

What both works reveal is the central role that risk perceptions play in the regulation of risk events. In the words of Ulrich Beck, "the *perception* of threatening risks determines thought and action" in modern risk societies.[49] As such, the following paradox has been created. The quest for safety against risk events which "transcend traditional boundaries of time and space" preoccupies the Western societies.[50] Yet, this quest is not necessarily guided by probabilistic risk analysis but is very much influenced by the (fears and) reactions of the public to (the threat of) risk events.

As such, the developer of a regulatory system addressing catastrophic risks is required to manage social perceptions, as much as real risks. This is not an easy

[45] R. Posner, *Catastrophe. Risk and response* (OUP 2004), pp. 28–29.

[46] Ibid., p. 250.

[47] P. Slovic, "Perception of risk from asteroid impact" in P. Bobrowsky and H. Rickman (eds.) *Comet/Asteroid impacts and human society* (Springer 2007), pp. 369, 379.

[48] Ibid.

[49] U Beck, *World risk society* (Polity Press 1999), p. 135.

[50] G. Mythen, *Ulrich Beck: A critical introduction to the risk society* (Pluto Press 2004), p. 22.

task, as it only takes the occurrence of a risk event to lead to, what Ulrich Beck describes as, the "the de-bounding of risk perceptions and fantasies".[51] A collision between a NEO and the Earth that causes death and property damage in a large scale is a first-rate opportunity for the media, provided Earth is saved from annihilation, to amplify its consequences by creating a (more or very) dramatic version of reality. Similar past events have not caused such reaction either because they took place before living memory or they were near-misses in the sense that they resulted in relatively limited damages – at least compared to what could have been.[52]

From the legal standpoint, the most important part of this amplification process will be the search for culpable parties among those who were in charge of managing such risks. One might argue that this search for culpability is a far-fetched scenario in the case of NEOs, as (i) a collision between an asteroid and the Earth cannot be attributed to technology[53] and (ii) the scientists dealing with their detection/mitigation seem to enjoy a high degree of credibility by the media: "apparently the media had treated their activities in a positive light and had not interpreted their public statements as particularly ill-founded or self-serving".[54]

However, history suggests that media are quick to change attitudes vis-à-vis scientists and risk managers in the aftermath of risk events. In their attempt to capitalise on the "impact factor" (no pun intended) of risk events, they construct them as the result of moral failures of risk managers.[55] Risk events, man-made or natural, are the result of their personal traits and mental processes, such as forgetfulness, inattention, poor motivation, carelessness, and recklessness.[56] For how long this perception lasts depends on the gravity of the risk event in question and the post-event reactions of the industry and States.

The fact that the collision of an asteroid with the Earth is a natural event is irrelevant to this construction. The more accurate the identification of NEOs and the avoidance of collisions become, the more vulnerable to a legal action the bodies in charge will be. As the actual risk creator is absent, the search for culpability moves to the human actors who allegedly facilitated the collision by missing the NEO or did not take the appropriate mitigating action as a result of a judgement error. The actions filed against National Oceanic and Atmospheric Administration (NOAA) in the aftermath of the Indian Ocean earthquake and tsunami in 2004, surprising and frivolous as they were, demonstrate the issue at hand.[57] The search

---

[51] U Beck, *supra* note 50, p. 44.

[52] See *supra* para. 1.2.

[53] R. Posner, *supra* note 46, p. 24.

[54] P. Slovic, *supra* note 48, p. 380.

[55] T. Horlick-Jones, "Modern disasters as outrage and betrayal" (1995) 13 International Journal of Mass Emergencies and Disasters, pp. 305, 311.

[56] J. Reason, "Human error. Models and management" (2000) 172 West J Med, pp. 393–396.

[57] For details on the history of the claims, see A. Ito, *Legal aspects of satellite remote sensing* (Martinus Nijhoff 2011), pp. 191–194 and University of Nebraska-Lincoln's Programme on Space and Telecommunications Law, "Legal Aspects of NEO Threat Response and Related Institutional Issues. Final Report" (9 February 2010), para. 4.3, online: Secure World Foundation http://

for culpability is not restricted to the risk creator but, in its absence, extends to those in charge of monitoring the risk situation and issuing warnings about imminent risk events.

As a result, any future regulation of NEOs should not be limited on their quantitative aspect but also embrace their qualitative elements in equal measure. This will only be achieved if the scheme has the following dual scope.

Firstly, it should facilitate the advancement of R&D on the identification and mitigation of the risks associated with NEOs. The overarching aim in this part is to satisfy the demand for collecting more data on the actual risks posed by NEOs, setting accurate early warning systems, identifying remedial actions at a global level, and creating emergency response protocols in cases of accidents. Currently, the scientific community has a much better understanding of the risks posed by NEOs compared to 1992 when the survey was published.[58] This development has been achieved despite the limited funding and the persisting doubts about the credibility of the threat posed by NEOs. The results demonstrate that the regulators cannot afford to ignore the said risks anymore. Not surprisingly, the international space community has started to discuss the potential and the form of a future international regulation. The creation of the IAWN and the SMPAG is a step to the right direction.[59] Yet, it shall be treated as the first step towards a comprehensive international regulation of NEO risks.

Secondly, it should deal with the legal aftermath of an accident. The overarching aim in this part is to create a system that contains the "de-bounding of risk perceptions" and its ensuing effects by separating culpability from compensation. It is to be hoped that this distinction would create incentives for co-operation and trust-building among the various actors involved. Any alternative would lead to adjudication and blameworthiness that will inevitably erode the fundamental aim of such system, namely, the effective protection against the risk of NEOs.

The debate over the structure of a scheme dealing with NEOs is a nascent one with many avenues to explore. Understandably, the spatial constraints of this publication do not permit an all-inclusive analysis. Yet, the aim of this part of the paper is to contribute to the debate by providing "food for thought" to the drafters of a future scheme.

10.4.2 Policy Considerations in Establishing the Fund

Very few States have the technical capability to manage the risks posed by NEOs. They lead the way in R&D, yet they are required to work within the framework of the fundamental principles of space law, namely, the use of outer space for the

swfound.org/media/40426/legal_aspects_neo_response_institutional_issues_final_report.pdf (last accessed on 15 January 2018).

[58] *Supra* Chap. 3.

[59] *Supra* para. 3.1.1.

benefit of humankind and in the interest of promoting international co-operation and understanding, taking into consideration the interests of all State-Parties to the Outer Space Treaty.[60] These principles call for the creation of a scheme that has the interests of humankind at its centre, ideally involving all States in the decision-making process, with arguably the most pressing issue being the establishment of a duty of "international co-operation and assistance on planetary defence".[61]

At the same time, the creation of a scheme dealing with NEOs at a global level would be a challenge for regulators from a risk management perspective unlike anything seen until today. The reason of this challenge is not necessarily the catastrophic potential of NEOs. It is the sui generis nature of such system that is required to have an all-inclusive risk management role: provide financial support for developing technology to identify and mitigate NEOs (precautionary risk management); decide on the appropriate course of action in the case of detecting, early enough, that an NEO is in a collision path with the Earth (proactive risk management); and deal with the aftermath of a collision in terms of employing the necessary emergency and investigation protocols, making funds available for funding the rescue efforts and compensating victims (reactive risk management).

Such a tripartite structure is not common in the international or national field. Most schemes dealing with risk events handle (part of) their reactive risk management function by acting in lieu or on top of the cover provided by commercial insurers. The common denominator behind the International Civil Aviation Compensation Fund (ICACF, aviation terrorism),[62] the International Oil Pollution Compensation Funds (IOPC Funds, oil pollution at sea),[63] the Criminal Injuries Compensation Scheme (CICS, violent crime in Great Britain), the Motor Insurers' Bureau (MIB, road accidents in the UK), and the schemes for the risks posed by nuclear power plants[64] is that they "socialise" the financial burden of relevant risk events. This is achieved (i) by creating a system where the various commercial actors involved in the relevant risk-generating activity (including consumers in ICACF and (indirectly)

[60] Arts I, III, and IV of the 1963 Treaty on Principles Governing the Activities of States in the Exploration and Use of Outer Space, including the Moon and Other Celestial Bodies.

[61] The analysis of this duty is outside the scope of this paper.

[62] The ICACF is set up by the 2009 Convention on Compensation for Damage to Third Parties, Resulting from Acts of Unlawful Interference Involving Aircraft. As of 4 August 2015, the convention has not entered into force.

[63] The Funds are set up by the 1992 International Convention on Civil Liability for Oil Pollution Damage, the 1992 International Convention on the Establishment of an International Fund for Compensation for Oil Pollution Damage, and the Protocol of 2003 to the 1992 International Convention on the Establishment of an International Fund for Compensation for Oil Pollution Damage.

[64] See the 1960 Paris Convention on Third Party Liability in the Field of Nuclear Energy, as amended by the Additional Protocol of 28th January 1964, by the Protocol of 16th November 1982, and the 2004 Protocol to amend the 1960 Paris Convention as amended. See also the 1963 Convention supplementary to the 1960 Paris Convention as amended. Finally see the 1963 Vienna Convention in Civil Liability for Nuclear Damage, as amended by the 1997 Protocol to amend the 1963 Vienna Convention and the 2007 Vienna Convention on Supplementary Compensation for Nuclear Damage.

in MIB) pay out admissible claims, most often but not necessarily, via a Fund set up for this purpose (ICACF, IOPC, MIB), (ii) by creating a clear obligation for States to foot the bill of the risk event on the basis of policy reasons (CICS), or (iii) by combining the two, creating a system where the contributions towards compensation are coming from both commercial actors and governments (nuclear power plant risks).

Understandably, one might wonder why such a "light" structure is not recommended in the case of NEOs.

One reason is political and has been one of the drawbacks of ICACF: depositing a few billion dollars in an international organisation to deal with the effects of a low-probability risk event does not appeal to contributors, especially in times of austerity. The most often utilised schemes either cover high(er)-probability risk events (IOPC funds, MIB), are set up and administered by domestic governments based on their policy choices (CICS), or support industries that are considered desirable despite their potential for catastrophe (nuclear power plants). In that respect, it is expected that the establishment of a compensatory-only scheme for NEOs will go nowhere. It will face reluctance from most governments to commit funds for dealing with the aftermath of a risk that might never materialise in the lifetime of their voters. This is so, of course, unless a catastrophic event involving NEOs takes place.

Secondly, the establishment of a compensation-only fund will address the arguably less relevant (at the moment) issue of recovery. That will leave in a state of flux the most pressing issues of funding, which are the loss prevention research and finding avenues of consensus in managing the relevant risks. At the same time, such choice will fail to lay the groundwork for creating a co-operative decision-making system, as it will be limited to (a small part of) the postaccident phase of NEO-related events. In essence, all that it will do is to postpone the creation of a framework that will reconcile issues of sovereignty with best practices of risk management formulated by the few technically capable States. Can this be interpreted as giving permission to the said states to proceed with their endeavours in a unilateral manner? Probably not; yet, it will certainly steadfastly ignore the elephant in the room.

Similarly, suggestions to create "light" structures by focusing on specific legal issues, such as the use of nuclear weapons to prevent collisions,[65] shall be dismissed. They contribute minimally to the main aim of identifying the risks and enhancing our defences at a global level. At the same time, they are, on their own, a "hard sell" to governments which feel that they have limited space for compromise when the negotiating agenda is so restricted.

In that respect, it is submitted that the most appropriate way forward is to set up a holistic scheme that will regulate all three risk management phases of NEO risks. Setting up such a scheme might take longer than opting for a fragmented, piecemeal framework. Yet, its creation is expected to satisfy all relevant interests, legal and

[65] See H. Mayer, "Is a special legal regime for planetary defence measures necessary?" (6 June 2015), online: International Academy of Astronautics http://iaaweb.org/iaa/Scientific%20Activity/conf/pdc2015/IAA-PDC-15-06-06po.pdf (last accessed on 15 January 2018).

operational, provided it is designed in a manner that promotes flexibility in decision-making. The flexibility is essential in order to counteract one of the main arguments against such grand design, namely, that it will impair the effectiveness of mitigating actions and emergency responses: "...states [with technical capabilities] might resent detailed interference with an actual mitigation campaign, and more likely still might not accept being ordered to act by an international body, even if it would be a Security Council of which they themselves might be members".[66]

However, this is a misplaced argument. The examples of organisations such as ICAO and, to a lesser extent, IMO demonstrate that the creation of international legal norms via international organisations can be beneficial to the development of global industries that rely on R&D. Admittedly, the said international organisations do not have as broad a mandate as the one suggested in this paper. Yet, they have proven that collective decision-making can add value to the regulated activity. In that respect, one should not overlook the practical and educational value of having an international forum dealing with the entire risk management process. Creating such a hub is expected to stimulate a multidisciplinary dialogue early in the process of drafting best practices and legal norms, making them as fit for purpose as possible. The importance of such practice cannot be overstated as it will be beneficial to the fast(er) development of our planetary defence system and will fit like a glove with the philosophy of communality of space.

What follows are a few ideas on the structure of a future scheme on the management of the risks posed by NEOs on the basis of our analysis so far.

10.4.3 Creation of an International Planetary Defence Organisation

It is recommended that the scheme is set up by means of an international treaty (the Treaty) that will be the constitution of an international organisation (the Organisation) with its own legal personality and system of governance. This way it will enjoy independence from existing schemes, its structure will reflect the idiosyncrasies of NEOs, and a message will be given that the international community takes the relevant risks seriously.

10.4.3.1 Structure

It is important that this organisation deals with both the quantitative and the qualitative aspects of the risk of NEOs. To achieve this, it is essential to have a structure that (i) facilitates collective decision-making even in periods of crisis, (ii) provides for the adoption of the necessary legal norms and best-practice protocols, and (iii) provides funding for research, prevention, and crisis management, including the settlement of claims.

---

[66] Nebraska legal report, *supra* note 58, para. 6.8.

It has been rightly suggested that the decision-making process at the highest level shall be based on the paradigm of the relationship between the UN Security Council and the UN General Assembly. This is so because "[i]n practice...the result will be individual or multilateral action by sovereign states acting within the confines of the mandate [given by the Council] but otherwise making their own decisions as to how to apply force".[67] Correct as this recommendation is, this relationship shall be replicated in the context of the new Organisation. In that respect the Organisation shall consist of a Council and an Assembly. The Council shall be entrusted with maintaining international safety against the threat of NEOs, and the Assembly shall give legitimacy to its decisions, keeping a watchful eye on the interests of mankind.

Nevertheless, the new scheme should not fit into the confines of the UN for the following two reasons. One reason that has already been advanced is that the replication will permit a restructuring of the membership of the new Council to represent the space capabilities of the State-Parties.[68] An additional reason can be identified: the creation of an independent organisation would facilitate the creation of legal norms and best practices focusing on the risk management of NEOs. This can be achieved by incorporating organs similar to the IAWN and SMPAG ("IAWN" and "SMPAG") into the structure of the Organisation, making them (with the necessary adjustments in terms of membership and mandate) the powerhouses of the Organisation. They can play the role of the Air Navigation Commission of ICAO where a group of 19 aeronautical experts are responsible for proposing to the ICAO Council amendments to the existing technical Annexes to the Chicago Convention, as well as the creation of new Annexes.[69] "IAWN" shall be focusing on the precautionary risk management phase and "SMPAG" on the proactive/reactive phases. They would consist of experts in their respective fields, and one of their aims would be to draft relevant legal norms and best practices. Such norms would be recommended for adoption to the Council, which upon approval would submit them to the State-Parties. The norms will be finally approved, if the required number of State Parties does not file objections against them within the time frame prescribed by the Treaty.[70]

It is important to note that the "IAWN" and "SMPAG" via subsidiary organs and national space agencies shall undertake a public awareness campaign about NEOs. Such a campaign shall inform the public about what NEOs are, the risks they pose to our life, and the various mitigation methods that can be used to prevent collisions. The long-term aim of such campaign would be to persuade the public to support the funding for managing the risks of NEOs and at the same time to educate it to see through the biases of media in reporting risk events.

[67] *Ibid.*, para. 6.7.

[68] Ibid., para. 6.13.

[69] Articles 56 and 57 of the Convention on International Civil Aviation, signed at Chicago, on 7 December 1944 (Chicago Convention).

[70] See also Article 90 of the Chicago Convention.

10.4.3.2 Contributions to the Fund

With respect to funding, it is important that the Treaty sets up a Fund ("the Fund"). The Fund shall be administered by the Secretariat that is headed by a Director. The Director shall be responsible for the management of the Fund and in particular its compliance with internal and external financial regulations. (S)he will also be in charge of all the functions of the Secretariat. The Secretariat will have a dual role. It will assist the Director in the management of the Fund by providing financial and legal expertise and at the same time provide administrative support to the Assembly and the Council.

For building up the capacity of the Fund, it is necessary that all State-Parties to the Treaty automatically become parties to the Fund and make the necessary financial contributions. The Fund shall not have a compensatory-only scope. Instead, it shall provide financial support to the State-Parties engaged in NEO-related research, cover the costs of mitigation campaigns taking place under the remit of the organisation, and be the exclusive source of compensation to victims of collisions, provided the loss occurs in the territory of a State-Party. For humanitarian reasons, the Treaty might make provision for non-State-Parties to use the Fund in case of damage in their territory, upon approval of the Assembly and provided they commit to becoming Parties to the Treaty.

It is recommended that the Treaty makes *no pre-allocation* of the funds to each of the three aims of the Fund. The funds could be allocated on a yearly basis by the Council to reflect the risk profile of NEOs and the state of research, always leaving a reserve to deal with a crisis. This decision of the Council shall be taken following a joint recommendation from the "IAWN" and the "SMPAG". This arrangement is expected to give the necessary flexibility to put the funds in the best possible use in terms of managing the risks of NEOs on any given year. It is suggested that a more flexible decision-making process shall be adopted on how to spend the allocated funds: the Assembly, having agreed on broad guidelines, shall delegate the decisions on which projects to be funded to the "IAWN" and "SMPAG". This delegation of authority is justified on the basis that the role of "IAWN" and "SMPAG" will be to have a hands-on role with respect to the three phases of NEO management. Still, an emergency reallocation procedure shall be provided, in case the funds of the reactive risk management phase are not sufficient to cover the exposure of the Organisation to a risk event occurring in the midst of the year.

The rate of contributions shall be decided by the Assembly. It can be based on a number of factors, such as the size of the country, its expenditures on research, and whether it provides means for mitigating NEOs at a global level. The idea behind this suggestion is that bigger countries have a higher probability of being hit by an asteroid. At the same time, countries with no or little technical capability might be requested to offer more funds to support the ones that play an active role in the management of the risks. Admittedly, this is an area that the Organisation shall keep an open mind: if commercial activities ever take place in space in mass, a "saving the humankind" levy might be imposed on their organisers.

10.4.3.3 The Function of the Fund

In case a NEO either requires the launch of a mitigation campaign or causes personal injuries or property damage in a State-Party, the role of the Fund would be to contribute towards the cost of the campaign and/or the compensation of the victims. In that respect, the following three issues shall be addressed:

International Mandate for Action

Firstly, it has been rightly argued that the sharing of such costs by the international community should depend on the mitigation campaign taking place "within a mandate established by a concurrence of General Assembly and Security Council actions";[71] without such mandate, the cost shall fall onto the State that took the unilateral action.[72] Harsh as this might sound, it is an important provision in order to establish the authority of the Organisation and make planetary defence a truly international attempt. Having said that, it is equally important that an emergency procedure is established to deal with situations that require immediate response leaving no time for seeking such mandate. That might provide for the urgent convening of the Council or for giving the mandate to "SMPAG" in advance to initiate the mitigation procedures subject to the urgent approval of the Council. An alternative already suggested is to adopt a procedure along the lines of the Emergency Special Sessions of the UN General Assembly.[73] This procedure shall become part of the Treaty, yet its effectiveness might be limited by the fact that only very few States have the capability to operate mitigation campaigns. Unless they are on board, there is little that can be done by the rest of the international community. However, the establishment of a duty of "international co-operation and assistance on planetary defence" has the potential to change the dynamics of this relationship.

Liability and Risk Management

Secondly, grave doubts have been expressed about the suitability of the 1971 Liability Convention and/or other existing international norms to deal with questions of State liability that might arise in the process of managing NEO risks.[74] The drafters of the Treaty shall take the opportunity to end once and for all this debate by waiving the liability of the States participating in the research of NEOs and/or undertaking the mitigation/emergency response to a risk event. Such a waiver shall be broad enough to cover both the failure to warn and the failure to adequately miti-

[71] Nebraska legal report, *supra* note 58, para. 6.9.

[72] Ibid., para. 6.10.

[73] Ibid., para. 6.6.

[74] Ibid., paras 4.1 ff.

gate the relevant NEO risk/manage the emergency response. The necessary condition for granting such immunity shall be for the States in question to operate within the remit of the Organisation and comply with the provisions of the Treaty. It has been suggested that the waiver shall be granted upon the additional condition that the response of the State in question was "a *bona fide* effort at NEO threat mitigation".[75] Identifying what constitutes "*bona fide* effort" shall be approached with caution, because it has the potential to invite litigation from the back door: it needs to be formulated clearly and precisely, avoiding legal terms, such as wilful misconduct or even reasonableness, that are subject to different interpretations from various legal systems.[76] Admittedly, a wide immunity will face the reluctance of some States. Yet, it is the best way forward in setting the basis for a holistic system of planetary defence that aims to serve the international community against a natural risk.

In essence, the Fund shall be responsible for settling the level of compensation paid to the victims, detaching the risk management process from questions of blameworthiness. The Treaty shall give exclusive jurisdiction to the Fund to settle claims in the aftermath of the risk event, with their negotiations performed by claim managers employed by the Fund as part of the "SMPAG". In that respect, particular care shall be taken when identifying the types of damages to be recovered. The drafters might opt for a conservative approach, limiting them to death, personal injury, and property damage that are the direct consequences of the risk event. A liberal approach, which is witnessed in recent treaties dealing with the risks of nuclear power plants, is to identify the instances of recoverable economic losses, such as economic loss arising from personal or property damage, including the cost of restoring the damaged property, the loss of income incurred as a result of the damage/destruction of the property, and the cost of taking measures following the collision to prevent further damage. Considering the catastrophic potential of collisions with NEOs, a limit on a per claim basis shall be set in the Treaty, especially if the drafters opt for a liberal definition of the term damage. Setting such a limit is important as it will assist the calculation of the contributions to the Fund and will make less likely a scenario where the Fund runs out of money before compensating the victims. At the same time, a provision giving priority to claims for personal injury over property damage shall be included in the eventuality that there are not enough funds to cover both of them.

Distinguishing culpability from recovery is important for an additional reason. It gives the message that "we are all in this together", contributing, as such, in the trust-building exercise and the normalisation of life that slowly takes place after every risk event. Litigation will inevitably slow down this process, as its adversarial

---

[75] Ibid., para. 4.29.

[76] For a review of the varying interpretations of the term wilful misconduct in private international law conventions, refer to M Clarke, *International Carriage of Goods by Road: CMR* (Informa Law from Routledge 2014), paras 100–103 and D Mclean et al.; *Shawcross and Beaumont on Air Law* (Lexis Nexis 2017), Chapter 30, paras 515–534.

characteristics will bring to the public nothing but disappointment and a sense of betrayal, especially if accompanied by delays in receiving the compensation.[77]

It would be impossible to create a liability system that is immune from litigation. Still, the opportunities to litigate in the aftermath of a NEO risk event shall be limited to the following two occasions: (i) the victims shall be given the right to bring an action against the Fund before the courts of the place where the damage took place in case the settlement negotiations fail and (ii) the Fund shall be given the right to recover from States (Parties or not) that acted unilaterally causing or facilitating damage that was covered by the Fund. It is important that the right of recourse against non-conforming States is retained by the Fund so that the victims are compensated without delay following the risk event and are not involved in a process that might require sensitive diplomatic negotiations. These (most probably) protracted and expensive attempts for recovery shall fall on an institution that has the means to deal with such eventuality.

Parallel Establishment of Domestic Funds

Thirdly, it is important that the Treaty permits domestic funds to be set up dealing with compensation. States might set these up if they are willing to give compensation to their residents on top of the limits provided by the Fund. However, the conditions under which these funds will provide compensation shall be aligned to the requirements of the Treaty, i.e. types of recoverable damages. Furthermore, the only function of these domestic funds would be to pay additional compensation to the victims without any rights of recourse left to the Fund to pursue.

10.5 Conclusion

In conclusion, NEOs represent a major, yet underestimated, threat to our planet and possibly to our future as species. The current efforts undertaken are inefficient, because they are not globally coordinated and have limited financial resources.

The most appropriate solution would be to create an international organisation that will be in charge of the three risk management phases of NEOs, i.e. financial support for R&D related to identification and mitigation of NEOs (precautionary risk management), decision-making mechanisms to face an imminent NEO threat (proactive risk management), and funding the rescue efforts and compensating victims (reactive risk management). This would be a sui generis construction that reflects the unique characteristics of NEOs.

The analysis in this paper is by no means exhaustive. It purports to contribute to the dialogue that (reluctantly) takes place on the future regulation of NEOs by arguing that the effectiveness of our planetary defences requires unique regulatory

[77] G Leloudas, *Risk and liability in air law* (Informa 2009), para. 8.9.

solutions. In that respect, it is suggested that the drafters of such scheme shall be prepared to break away from traditional methods of international regulation which limit themselves on mitigating the effects of a risk event rather than preventing its occurrence. The catastrophic potential of the risk that we are dealing with leaves no other options.

Chapter 11
Legal Challenges of the New Space Race to Mars: Proposal for the Use of a Three-Tier Legal Framework

Sarah Germann and Anja Nakarada Pecujlic

11.1 Introduction

11.1.1 The New Space Race to Mars

After a lengthy period of declining national space activities and manned missions to space, and following a stalemate to the evolution of international space law, the international space community has recently reached a turning point.[1] With the influx of the private sector as a new player in space exploration, a multifaceted new space race has begun in terms of dominance over the space sector. An ongoing competition on who will be the first to deploy another manned mission to the Moon and a manned mission to Mars or even to establish a first Martian settlement seems to be real.[2] This race takes place among governments, private ventures and, increasingly, between private and public space actors. In addition, although it is uncertain whether private ventures will manage to reach Mars before governmental entities, the biggest part of the necessary technology will, undoubtedly, be

[1] See further Set of Questions provided by the Chair of the UN COPUOS Legal Subcommittee's Working Group on the Status and Application of the Five UN Treaties on Outer Space, taking into account the UNISPACE+50 process.

[2] Boeing CEO Says Boeing Will Beat SpaceX to Mars, Space.com, https://www.space.com/39014-will-boeing-beat-spacex-to-mars.html, last accessed 19.03.2018; The Race to Mars: Here's How SpaceX Ranks against the Competition, The Verge, https://www.theverge.com/2016/9/30/13114704/spacex-elon-musk-vs-mars-one-nasa-mission-timeline, last accessed 19.03.2018; Journey to Mars Overview, NASA, Washington DC, USA, https://www.nasa.gov/content/journey-to-mars-overview, last accessed 19.03.2018.

S. Germann (✉)
Department of International Law in the Office of the Legal Adviser,
Austrian Foreign Ministry, Vienna, Austria

A. Nakarada Pecujlic
University of Cologne, Cologne, Germany

G. D. Kyriakopoulos, M. Manoli (eds.), *The Space Treaties at Crossroads*,
https://doi.org/10.1007/978-3-030-01479-7_11

developed by the private sector.[3] Therefore, it can be argued that the private sector is currently becoming the driving force in space exploration.

Noteworthy are several examples of major activities conducted by private actors: the development of SpaceX rockets by Elon Musk and the most recent launch of the Falcon Heavy Rocket in February 2018, where a Tesla car was sent into Outer Space; Amazon founder Jeff Bezos' space tourism aspirations with his commercial space flight company Blue Origin, who in December 2017 sent a suborbital New Shepard rocket into space; Microsoft's Paul Allen and his space cargo delivery system called Stratolaunch Systems; and the US real estate mogul Robert Bigelow, who spent more than $200 million for commercial space station modules.[4] Similarly, a group, which also comprises Google executives Larry Page and Eric Schmidt, formed an asteroid mining company called Planetary Resources which sets out to exploit the natural resources to be found in asteroids by 2020.[5]

One of those aspiring projects is the Mars One Project by entrepreneur Bas Lansdorp, which received broad media coverage and caught the world's attention. Mars One started off in 2011 as a Dutch-based non-profit foundation, which publicly announced its plans to establish a permanent human settlement on Mars made out of volunteers.[6] The deployment of the first manned mission is currently planned for the year 2031, with a prospective landing in 2032.[7] Originally, the 6-billion dollar venture was set out to be funded through media sponsorships and an intended reality TV show, as well as by 94 countries worldwide.[8] Hence, the Mars One project was drafted as a non-governmental multi-national collaboration: the project would be independent but funded internationally.[9]

---

[3] See, e.g. "Public-private partnerships are the future of space exploration", statement from Dava Newman, a former NASA deputy administrator, CNBC News, February 2018; Six Private Companies That Could Launch Humans into Space, Space.com, https://www.space.com/8541-6-private-companies-launch-humans-space.html, last accessed 19.03.2018.

[4] E. Helmore, Who owns the moon? "Space lawyers" increasingly needed for legal issues beyond Earth's atmosphere, in The Telegraph, Sept 26, 2013, at http://news.nationalpost.com/news/who-owns-the-moon-space-lawyers-increasingly-needed-for-legal-issues-beyond-earths-atmosphere; see further Start-Up Space 2018: Update on Investment in Commercial Space Ventures, A Bryce Space and Technology Publication, 2018.

[5] S. Worral, Three billionaires are racing for space. Who will win?, National Geographic, Apr 7, 2018, at https://news.nationalgeographic.com/2018/04/three-billionaires-are-racing-to-space%2D%2Dwho-will-win-/; Planetary Resources, About the Exploration Programme, at https://www.planetaryresources.com/missions/arkyd-301/; E. Helmore, Who owns the moon? "Space lawyers" increasingly needed for legal issues beyond Earth's atmosphere, in The Telegraph, Sept 26, 2013, at http://news.nationalpost.com/news/who-owns-the-moon-space-lawyers-increasingly-needed-for-legal-issues-beyond-earths-atmosphere.

[6] M. Listner and C. Newman, Failure to launch: the technical, ethical, and legal case against Mars One, in The Space Review, Mar 16, 2015, at http://www.thespacereview.com/article/2712/1.

[7] Mars One, Mars One presents an updated mission roadmap, Jan 31, 2017, at http://www.mars-one.com/news/press-releases/mars-one-presents-an-updated-mission-roadmap.

[8] Madeleine Stern, Mankind is Mars-Bound: All the Facts on Mars One, June 11, 2014 at https://lawstreetmedia.com/issues/health-science/mankind-mars-bound-facts-mars-one/.

[9] Madeleine Stern, Mankind is Mars-Bound: All the Facts on Mars One, June 11, 2014 at https://lawstreetmedia.com/issues/health-science/mankind-mars-bound-facts-mars-one/.

More recently, the entire international community became aware of a more ambitious undertaking: Elon Musk's plan to "colonize Mars" with his venture SpaceX. Indeed, in September 2017, at the 68th annual International Astronautical Congress, in Adelaide, Australia, he announced his plan to build BFR rockets that will be able to support long-duration spaceflights,[10] and a few months prior to the announcement, he published his rather detailed plan to build a civilization on the Red Planet, titled "Make Humans a Multi-Planetary Species".[11] His plan envisions enormous, reusable rockets, carbon fibre fuel tanks, ultra-powered engines, spaceships that can carry up to hundred people and the involvement of multiple investors.[12]

The aforementioned examples and many other innovative near-future plans for manned missions into deep space focus mostly on exploration, technology and innovation challenges, but none of them address sufficiently the legal implications of such endeavours.

One of the first problems that arise is the lack of awareness on international space law and its applicability to space activities on the part of the private pace sector.[13] Space entrepreneurs and their legal counsels tend to interpret established legal norms to their favour. According to them, space law is not applicable to individuals, while Mars should rather not be interpreted to qualify the definition of a celestial body.[14] As a result, the applicability of the international space treaties as well as established principles of space law is challenged. Especially as exploration might soon give way to settlement, it is important that settlement activities are not operated as long as a legal vacuum still exists. As it cannot be denied that space law was originally drawn up for state civilian or military actors, many are the uncertainties around the legal applicability of certain space law principles to planetary settlement.[15] Such uncertainties constitute destabilizing factors that could incite private actors and possibly their respective nations to act unilaterally and in violation of international law.

Secondly, complex and expensive space missions are usually realized through joint undertakings among different states and actors.[16] States opt to cooperate in order to share risks and expenses and in order to optimize expertise. In addition,

---

[10] BFR and BFR Overview, http://www.spacex.com/mars, last accessed 13.04.2018.

[11] E. Musk, New Space. June 2017, vol. 5, issue 2, pp. 46–61. https://doi.org/10.1089/space.2017.29009.emu.

[12] E. Musk, New Space. June 2017, vol. 5, issue 2, pp. 46–61. https://doi.org/10.1089/space.2017.29009.emu.

[13] M. J. List, International space law and commercial space activities: the rules do apply, in The Space Review, Jun 3rd, 2013, at http://www.thespacereview.com/article/2305/1.

[14] Amanda Wills, Is Mars for Sale?, in Mashable, Apr 09, 2013, at http://mashable.com/2013/04/09/mars-land-ownership-colonization/.

[15] E. Helmore, Who owns the moon? 'Space lawyers' increasingly needed for legal issues beyond Earth's atmosphere, in The Telegraph, Sept 26, 2013, at http://news.nationalpost.com/news/who-owns-the-moon-space-lawyers-increasingly-needed-for-legal-issues-beyond-earths-atmosphere.

[16] J.Hurtak, M. Egan, "A Multilateral Agreement For A Manned Mission To Mars" p 2 (2006).

private actors are involved as contractors and subcontractors to develop necessary technology in a more timely and costly manner. For example, ESA, Roscosmos and private companies originating from more than 20 countries collaborated in the ExoMars mission, with ESA building the orbiter/lander and Roscosmos providing the launch vehicle.[17] Nevertheless, whenever there is no overarching governmental project with precise norms regulating the rights and obligations of all the actors involved, legal uncertainties arise, and further development of technology and ground-breaking space missions is hindered.

This is mainly due to the responsibility mechanism, which is comprised in space law, according to which states are internationally responsible for governmental and non-governmental activities.[18] Accordingly, states might be reluctant to issue the required national licenses for private actors to develop new technologies in the event of private ventures with independent space endeavours. This, in turn, leads to the private sector not being incentivized to invest, as it is uncertain if benefits be derived from these costly ventures. However, the encouragement of private investment is necessary to develop new technologies and, generally, to progress. In this regard, clarity in legal provisions is essential for states to be aware of their responsibilities so that they can issue the necessary licenses to private companies and, as a result, contribute to the future of technological advancement even when no government-initiated projects are involved.

Therefore, the problem that arises is twofold: on one hand, orderly space exploration will be endangered in case unilateral decision on the applicability of space law principles takes place by the private space sector; and, on the other hand, necessary technological development will be delayed due to existing legal uncertainties. To address both challenges, it is most urgent, that the existing legal gaps be clarified and filled. As mentioned earlier, the amendment of the existing treaties or the creation of a novel space treaty is unlikely due to geopolitical reasons, hence the need for an innovative solution.

11.1.2 Brief Summary of the Existing Legal Framework

The history of outer space law begins with the United Nations General Assembly's ad hoc Committee on the Peaceful Uses of Outer Space (UNCOPUOS), which, in 1958, addressed the nature of the legal problems that could arise from space activities.[19] As a result, the Declaration of Legal Principles Governing the Activities of

[17] Robotic Exploration of Mars, ExoMars Mission team, ESA, Paris, France, http://exploration.esa.int/mars/56623-exomars-mission-team/, last accessed 18 January 2018.

[18] Art VI, Treaty on Principles Governing the Activities of States in the Exploration and Use of Outer Space, including the Moon and Other Celestial Bodies, 610 UNTS 205, entry into force Oct 10, 1967.

[19] United Nations General Assembly. Resolution 1348 (XIII) (13 December 1958) "Question of the Peaceful Use of Outer Space".

States in the Exploration and Use of Outer Space[20] was adopted by the General Assembly in 1963 and was the predecessor for the OST.[21] The OST is the basis for all space activities, as it constitutes the foundation for the four following, more specific space treaties: the Agreement on the Rescue of Astronauts, the Return of Astronauts and the Return of Objects Launched Into Outer Space,[22] Convention on International Liability for Damage Caused by Space Objects,[23] Convention on Registration of Objects Launched into Outer Space,[24] and Agreement governing the Activities of States on the Moon and Other Celestial Bodies.[25]

The OST was based on the Antarctic Treaty.[26] The drafters of the OST draw similarities between the two areas, both being without human inhabitants, free for exploration and scientific investigation and purposed for exclusively peaceful use.[27] However, nowadays, the situation tends to be different as the first steps to inhabit outer space have already taken place.[28] Because the predominantly international body of space law was originally designed to cover the needs of civilian or military state actors and the OST was adopted in an era when the involvement of private space actors was not foreseen to its real extent,[29] law needs to adapt to modern needs and challenges. This is, because planetary civilization modelled after life on Earth, would give rise to a multitude of legal questions and might even seem incompatible with the parameters of the current framework of Space Law. The principle of nation-state, sovereignty and/or private ownership of immovable property, and the exploitation of natural resources for the preservation of Earth's population constitute the basis and the fundament of our "civilization" as we know it. However, all of these

---

[20] 1721 GA Resolution (XVI), V. Kopal, History of OST", http://legal.un.org/avl/ha/tos/tos.html.

[21] United Nations General Assembly. Resolution 2222 (XXI) of 19 December 1966 "Treaty on Principles Governing the Activities of States in the Exploration and Use of Outer Space, Including the Moon and Other Celestial Bodies".

[22] *Agreement on the Rescue of Astronauts, the Return of Astronauts and the Return of Objects Launched Into Outer Space*, 22 April 1968, 672 UNTS 119, 19 UST 7570, TIAS No 6599, 7 ILM 151 (entered into force 3 December 1968) [Rescue Agreement].

[23] *Convention on International Liability for Damage Caused by Space Objects*, 29 March 1972, 961 UNTS 187, 24 UST 2389, 10 ILM 965 (1971) (entered into force 1 September 1972) [Liability Convention].

[24] *Convention on Registration of Objects Launched into Outer Space*, 6 June 1975, 28 UST 695, 1023 UNTS 15 (entered into force 15 September 1976) [Registration Convention].

[25] *Agreement governing the Activities of States on the Moon and Other Celestial Bodies*, 5 December 1979, 1363 UNTS 3 (entered into force 11 July 1984) [Moon Agreement].

[26] Antarctic Treaty, 12 UST 794; 402 UNTS 71; 19 ILM 860 (1980).

[27] See further J. Hanessian, "The Antarctic Treaty 1959", The International and Comparative Law Quarterly, vol. 9 (1960).

[28] Journey to Mars Overview, NASA, Washington, DC, USA, https://www.nasa.gov/content/journey-to-mars-overview, last accessed 19.03.2018.

[29] P. R. Harris, Space Law and Space Resources, at http://www.nss.org/settlement/nasa/spaceresvol4/spacelaw.html; E. Helmore, Who owns the moon? 'Space lawyers' increasingly needed for legal issues beyond Earth's atmosphere, in The Telegraph, Sept 26, 2013, at http://news.nationalpost.com/news/who-owns-the-moon-space-lawyers-increasingly-needed-for-legal-issues-beyond-earths-atmosphere.

principles seem to clash with the fundaments of the current body of international space law, as in, for example, the non-appropriation principle and the absence of provisions regarding exploitation of natural resources.[30]

The following part will address the main principles of outer space law with possible impact on future manned missions to Mars and other extraterrestrial space settlements together with the relevant views adopted by the private space sector.

11.2 Legal Issues

The main legal issues arising from manned missions to future settlements on Mars comprise the non-appropriation principle[31] and the controversial issue of whether utilization of the natural resources of the Red Planet is allowed. The lack of clarity in such issues renders the establishment of civilian structures on extraterrestrial planets almost impossible. Therefore, these two principles – as well as their interpretation by the private space sector – require further investigation, together with issues such as the relevant liability regime, transfer of technology, information sharing, environmental management and import/export restrictions regarding space technology.

The example of the Mars One Venture will be examined to illustrate the complexity of the idea of a manned mission and subsequent settlement on the Red Planet and the legal challenges that arise from it.

11.2.1 Appropriation of Planetary Realty

Claims of sovereignty over celestial bodies or portions of geostationary orbits by numerous states and individuals are not new. However, those claims have always been successfully rejected by the major space faring nations as violating the existing international legal framework.[32] Article II of the (OST), the non-appropriation principle, stipulates: "Outer space, including the moon and other celestial bodies, is not subject to national appropriation by claim of sovereignty, by means of use or occupation, or by any other means".[33]

---

[30] F. Tronchetti, The Exploitation of Natural Resources of the Moon and other Celestial Bodies: A Proposal for a Legal Regime, p 3 (2009).

[31] Art II of the Outer Space Treaty provides as follows: "Outer space, including the Moon and other celestial bodies, is not subject to national appropriation by claim of sovereignty, by means of use or occupation, or by any other means".

[32] H. DeSaussure, The Freedoms of Outer Space and their Maritime Antecedents in Space Law: Development and Scope, p 3 (ed. N. Jasentuliyana, 1992).

[33] Article II, Treaty on Principles Governing the Activities of States in the Exploration and Use of Outer Space, including the Moon and Other Celestial Bodies, 610 UNTS 205, entry into force Oct 10, 1967.

The private sector's major counter-argument against the essence of the non-appropriation principle is that Article II OST only prohibits *national* appropriation. Accordingly, it has been noted that "international law bans governments from owning land on the Moon [or Mars], but private entities could legally own such land".[34]

However, the majority of scholars share the opinion that Article II prohibits private property of outer space in its entirety or celestial bodies or parts thereof, be it claimed or by states or by private entities.[35] Taking into account the Vienna Convention on the Law of Treaties (VCLT)[36] and interpreting the treaty in good faith and in its context, as well as acknowledging that the OST was drafted in the 1960s, it must be noted that the drafters could not have considered possible future private entities acting independently in space activities. Therefore, Article II OST only prohibits national appropriation *expressis verbis*. Regarding a possible private appropriation, this could lead to two different conclusions: either, taking into account the intent and rationale of the OST, private actors are implicitly covered by Article II OST as private ownership over celestial lands would run counter to the spirit of the OST; or, based on an *argumentum a maiore ad minus* approach, private ownership claims are prohibited, since neither are governmental ownership claims.

A later effort to prohibit private appropriation of space and celestial bodies exists in the Moon Agreement, which, however, does not enjoy wide acceptance due to the small number of its signatories.[37]

One more argument against private ownership of parts of outer space can be drawn from Article VI of the OST, according to which states are responsible for all their national activities, whether governmental or non-governmental, including activities on celestial bodies, and are required to ensure that the activities are carried out in compliance with the obligations of the treaty itself.[38] On the basis of Article VI, private governmental and non-governmental activities qualify as "national activities", able to constitute "national appropriation" through respective actions.[39] In addition, because states are required to authorize and supervise national

[34]The Space Settlement Institute, Lunar Land Claims Recognition, at http://www.space-settlement-institute.org/strategy.html.

[35]P. M. Sterns and L. I. Tennen, Privatering and Profitering on the Moon and Other Celestial Bodies: Debunking the Myth of Property Rights in Space, Proceeding of the Forty-Fifth Colloquium on the Law of Outer Space, p 60 (2002); F. Tronchetti, The Exploitation of Natural Resources of the Moon and other Celestial Bodies: A Proposal for a Legal Regime, p 29 (2009).

[36]Article 31, Vienna Convention on the Law of Treaties, 1155 UNTS 331, entry into force Jan 27, 1980.

[37]Currently only 17 states have ratified the Agreement. See Status of International Agreements relating to activities in outer space as at 1 January 2017, A/AC.105/C.2/2017/CRP.7.

[38]Article VI, Treaty on Principles Governing the Activities of States in the Exploration and Use of Outer Space, including the Moon and Other Celestial Bodies, 610 UNTS 205, entry into force 1967-10-10.

[39]Statement, Board of Directors of the International Institute of Space Law (IISL) on Claims to Property Rights Regarding the Moon and Other Celestial Bodies, at http://www.iislweb.org/docs/IISL_Outer_Space_Treaty_Statement.pdf.

activities, by doing so in the event of private ownership of celestial bodies, they would enforce such ownership through their national administrative mechanisms, and, as a result, they would be in breach of Article II of the OST.[40] Moreover, as a final argument, if a state is prohibited from engaging in a certain activity, it also lacks the right to authorize a private entity to undertake the same activity, based on the basic rule *nemo plus iuris*.[41] A number of scholars support a different position based on the "legal loophole" argument, according to which the lack of explicit prohibition in Article II excludes private ownership from its scope.[42] This approach is repeatedly advocated by the private sector in the benefit of their ventures.[43]

To sum up, the discussion itself illustrates the uncertainty to which the current regime could lead. Such uncertainty could result to further controversial outcomes. For instance, several opinions have already been expressed supporting state withdrawal from the OST under its Article XVI, based on the position that the current situation could not facilitate private ownership of outer space parts.[44] Such reactions, however, could have a serious impact and endanger the future of space activities.

11.2.2 Exploitation of Natural Resources Necessary for Establishment of Planetary Settlement[45]

Both the issue of private ownership or appropriation of planetary land and that of the utilization space resources have been approached from multiple angles. As already mentioned, the private sector tends to interpret – or even "bend" – the law

[40] R. Simberg, Homesteading the Final Frontier – A Practical Proposal for Securing Property Rights in Space, p 8–9 (2012); F. Tronchetti, The Exploitation of Natural Resources of the Moon and other Celestial Bodies: A Proposal für a Legal Regime, p 29 (2009).

[41] F. Tronchetti, Legal Aspects of Space Resource Utilization, in Handbook of Space Law (ed. Frans von der Dunk, 2015), p 781; F. Tronchetti, The Exploitation of Natural Resources of the Moon and other Celestial Bodies: A Proposal for a Legal Regime, p 31 (2009).

[42] S. Gorove, "Interpreting Article II of the OST", Fordham Law Review Vol. 37 No. 37, pp. 349–351 (1969).

[43] See, e.g. A. Wills, Is Mars for Sale?, in Mashable, Apr 09, 2013, at http://mashable.com/2013/04/09/mars-land-ownership-colonization/, last accessed 12.01.2018; Amanda Wills, Is Mars for Sale?, in Mashable, Apr 09, 2013, at http://mashable.com/2013/04/09/mars-land-ownership-colonization/, last accessed 02.02.2018; Lunar Embassy, Extraterrestrial Property and Space Law: Fact and Fiction, 1329 US Highway 395 North, Nevada, USA, http://lunarembassy.com/current-space-law/, last accessed 19.03.2018.

[44] R. Simberg, Homesteading the Final Frontier – A Practical Proposal for Securing Property Rights in Space, p 9 (2012).

[45] For example: water, carbon dioxide for oxygen and rocket fuel; nitrogen and other elements for fertilizers; iron oxide and silicon oxide for iron, steel and glass; geothermal energy as a power source. See M. Wall, Incredible Technology: How to Live on Mars (2013), at https://www.space.com/22342-how-to-live-on-mars-colony-technology.html; NASA, in Situ Resource Utilization (2012), at https://isru.nasa.gov/SPACERESOURCES.html.

in their favour: "It is allowed to use land, just not say that you own it. It is also allowed to use resources that you need for your mission", stated Bas Lansdorp in an interview on the topic of the Mars One Venture.[46]

From a legal point of view, the matter is more complex. One could refer to the remarks concerning Article II, as analysed earlier; however, the question whether the non-appropriation principle includes the exploitation of natural resources constitutes a separate discussion.[47] In general, it can be concluded that the vagueness of the provisions of the OST, in what it concerns the exploitation of the natural resources of the celestial bodies, renders the Treaty silent on the topic.

The Moon Agreement constitutes *lex specialis* to the OST as it is a subsequent treaty to it and is also applicable to "other celestial bodies within the solar system".[48] The *travaux préparatoires* of the Treaty show that its strict provisions hindered the wide signature and ratification of the Treaty.[49] Interestingly enough, the Netherlands is one of the signatory states to the Moon Agreement and the OST. This means that it is bound by the Treaties' provisions, and through Article VI of the OST, its private entities are so too.[50] Therefore, the Mars One Venture, as a non-governmental entity that is incorporated as a non-profit organization in the Netherlands, is subject to the continuing jurisdiction of the said state[51] and should thus act in conformity with the Moon Agreement. Therefore, the Moon Agreement shall be discussed in this paper.

Article 11 (3) of the Moon Agreement states that "neither the surface nor the subsurface of the moon nor any part thereof or natural resources in place, shall become property of any state, international intergovernmental or non-governmental organization, national organization or non-governmental entity or of any natural person". This means that neither the Netherlands, nor Mars One, could own, trade or use Mars' natural resources. On the contrary, the private sector has argued that, based on an interpretation of the wording of this Article and considering the literal meaning of the phrase "in place", an entity could own the extracted materials after their removal and processing through personal labour.[52] Arguably, ownership of

---

[46] A. Wills, Is Mars for Sale?, in Mashable, Apr 09, 2013, at http://mashable.com/2013/04/09/mars-land-ownership-colonization/.

[47] F. Tronchetti, The Exploitation of Natural Resources of the Moon and other Celestial Bodies: A Proposal for a Legal Regime, p 31 (2009).

[48] Article 1, Agreement Governing the Activities of States on the Moon and Other Celestial Bodies, 1363 UNTS 3, entry into force Jul 11, 1984.

[49] F. Tronchetti, Legal Aspects of Space Resource Utilization, in Handbook of Space Law (ed. F. von der Dunk, 2015), p 789.

[50] M. Listner and Christopher Newman, Failure to launch: the technical, ethical, and legal case against Mars One, Mar 16, 2015, in The Space Review, at http://www.thespacereview.com/article/2712/1.

[51] M. Listner and Christopher Newman, Failure to launch: the technical, ethical, and legal case against Mars One, Mar 16, 2015, in The Space Review, at http://www.thespacereview.com/article/2712/1.

[52] P. R. Harris, Space Law and Space Resources, at http://www.nss.org/settlement/nasa/spaceresvol4/spacelaw.html.

natural resources mined from the lunar surface has been recognized in the past[53]: lunar samples have been exchanged by the USA as well as by the Soviets for other tokens of value.[54]

As one can see, the space treaties provide no clarity when it comes to questions on the utilization of natural resources. Therefore, alternative regimes, such as an international investment organization, to which nations could become members, are needed. Intelsat, for instance, constitutes such a model.[55] Some scholars such as Karen Cramer of the Space Policy Institute in Washington have suggested the solution of Users Unions, based on the ITU.[56] Such unions would only grant rights of use to private enterprises and states for commercial mining and exploration, and their main function would be to ensure non-interference among groups wishing to pursue similar interests.[57] These, or similar, regimes need to be established always, however, guided by the philosophies of the OST and the Moon Agreement, in an efficient and universally binding manner.

11.2.3 Liability

The possibility of liability for damage is inherent in any space mission. Liability can be established for property damages and, in the event of manned missions, for injury to or loss of human beings.[58] Statistics show that 3.4% of all humans who have travelled into space lost their lives during a mission.[59] What is more, the fact that outer space is a majorly unexplored and untrodden environment renders the possibility of return uncertain as far as manned missions are concerned. Therefore, such missions can be characterized as extra-hazardous with a high risk of liability

[53] (Art 6 (2) Moon Agreement allows State parties the right to collect on and remove from the Moon samples of its mineral and other substances in carrying out scientific investigation).

[54] R. Simberg, "Property Rights in Space", in The New Atlantis, No 37, Fall 2012, pp. 20–31, at http://www.thenewatlantis.com/publications/property-rights-in-space.

[55] P. R. Harris, Space Law and Space Resources, at http://www.nss.org/settlement/nasa/spaceresvol4/spacelaw.html.

[56] K. Cramer, "The Lunar Users Union – An Organization to Grant Land Use Rights on the Moon in Accordance with the OST," IISL-97-4.13 Proceedings from the 40th Colloquium on The Law of Outer Space, Oct 5–10, 1997, p. 352; J. Hurtak, Legislation and Space Law Concepts proposed for the eventual industrialization of Mars by Men, p 860 (1998).

[57] K. Cramer, "The Lunar Users Union – An Organization to Grant Land Use Rights on the Moon in Accordance with the OST", IISL-97-4.13 Proceedings from the 40th Colloquium on The Law of Outer Space, Oct 5–10, 1997, p. 352.; J. Hurtak, Legislation and Space Law Concepts proposed for the eventual industrialization of Mars by Men, p 860 (1998).

[58] Article I, Liability Convention; I. Diedriks-Verschoor, An Introduction to Space Law, p 37 (1999).

[59] J. Silver, Houston, we have a (Liability) Problem, 112 Mich. L. Rev. 833 (2014).

for the responsible states (i.e. the launching states), although the latter neither take part in the ventures, nor do they, in any manner, benefit from them.

According to Article VII of the OST, the launching state "is internationally liable for damage to another state party to the treaty or to its natural or judicial persons".[60] As such, states are the only entities that can incur international liability,[61] even for missions conducted by private, non-governmental entities.

However, manned missions to Mars are complex as three or more states might simultaneously be the launching entities, depending on the location of the launch, the launching facilities used and the form of cooperation with other entities. As a result, there is a range of possibly liable states, participating space agencies and private actors. Accordingly, the problem can be extended to contractors and manufacturers that provide products and services to launch operators.[62] However, these actors play a critical role in every mission, incite progress and, therefore, need to be protected. In reality the challenges that these actors face do not only include *ex nunc* liability but also claims of insurance.[63] To mitigate this situation, incentives to attract the private space sector need to be established, together with risk mitigation mechanisms for states.

The aforementioned problems illustrate the ineffectiveness of the space treaties to address modern realities. Although both the OST and the Liability Convention provide the ground principles with regard to liability, specialized and detailed frameworks for individualized missions are essential.

11.2.4 Transfer of Technology and Information Sharing

Because of the level of technological development and the stage of respective space missions at the time when the five space treaties were created, no provisions concerning transfer of technology or information sharing were explicitly included in their texts.[64] However, several articles of the said regime can be viewed as implicitly including a duty to technology and information sharing, such as Article I of the OST, which includes a non-discrimination clause[65]; Article IX of the OST, which

---

[60] Article VII, Treaty on Principles Governing the Activities of States in the Exploration and Use of Outer Space, including the Moon and Other Celestial Bodies, 610 UNTS 205, entry into force Oct 10, 1967.

[61] V. Kayser, Launching Space Objects: Issues of Liability and Future Prospects, p 34–36 (2006).

[62] J. Silver, Houston, we have a (Liability) Problem, 112 Mich. L. Rev. 833 (2014).

[63] J. Silver, Houston, we have a (Liability) Problem, 112 Mich. L. Rev. 833 ff (2014).

[64] M. Mineiro, Space Technology Export Controls and International Cooperation in Outer Space, p158 (2011).

[65] Article I Outer Space Treaty states: "Outer Space (...) shall be free for exploration and use by all States without discrimination of any kind".

includes a due regard principle[66]; and Article IV of the Moon Agreement, which establishes the obligation for cooperation in space activities.[67]

The practice up to this time, however, shows that in this case, law does not reflect reality. Michael Mineiro notes that "State practice confirms that complete discrimination with regards to the sale, purchase, exchange and trade in civil space technologies and spacecraft is legal".[68]

In addition, although general international regimes for intellectual property exist, such as the TRIPS,[69] they have so far only achieved a certain level of harmonization among national intellectual property (IP) legislations, whereas there are still considerable matters solely regulated by national laws. These national regimes differ significantly from each other, which results in the necessity for a joint venture to customize the application of intellectual property rights individually within the different jurisdictions involved.[70] In addition, the manner in which national IP applies to extraterritorial activities is uncertain. This could lead to a legal fragmentation of multilateral ventures and as such to a patchwork of applicable norms contributing thus to further legal uncertainty.[71]

However, legal certainty is one of the most important factors that attract investment into the space industry. Especially, given the trend towards public-private joint ventures and missions, a safe and secure technology sharing and transfer regime is increasingly necessary.[72] More importantly, whenever entrepreneurs need to cooperate with or compete against governmental players or competitors who are subsidized by the state, uncertainties or risks will hinder further progress. Clear rules in the application of intellectual property law, technology transfer and information sharing, tailored to the specificities of space activities, are essential to give private

---

[66] Article IX Outer Space Treaty states: "In the exploration and use of outer space (…) State Parties to the Treaty (…) shall conduct all their activities in outer space (…) with due regard to the corresponding interests of all other State Parties to the Treaty".

[67] Article 4 para 2, Moon Agreement states: "State Parties shall be guided by the principle of co-operation and mutual assistance in all their activities concerning the exploration and use of the moon. (...)".

[68] M. Mineiro, Space Technology Export Controls and International Cooperation in Outer Space, p158 (2011).

[69] The TRIPS Agreement, which came into effect on 1 January 1995, is to date the most comprehensive multilateral agreement on intellectual property. It covers copyright and related rights, trademarks including service marks, geographical indications including appellations of origin, industrial designs, patents and undisclosed information including trade secrets and test data; see also https://www.wto.org/english/tratop_e/trips_e/intel2_e.htm; Agreement on Trade-Related Aspects of Intellectual Property Rights (TRIPS) Agreement Establishing the World Trade Organization, Annex 1C, 1869 U.N.T.S. 299, 33 I.L.M. 1197 (1994) [hereinafter TRIPS Agreement].

[70] WIPO, International Bureau, Intellectual Property and Space Activities, at http://www.wipo.int/export/sites/www/patent-law/en/developments/pdf/ip_space.pdf.

[71] WIPO, International Bureau, Intellectual Property and Space Activities, at http://www.wipo.int/export/sites/www/patent-law/en/developments/pdf/ip_space.pdf.

[72] J. Hurtak, M. Egan, "A MULTILATERAL AGREEMENT FOR A MANNED MISSION TO MARS" p 1 (2006).

industry the necessary confidence to progress in technology development.[73] Due to the complexity of the missions, a case-to-case regulation adapted to the needs of individual missions would be most preferable.

Based on the aforementioned, it results that there is a number of crucial legal loopholes under the current legal framework, which, on one hand, can lead to dangerous unilateral private undertakings and, on the other hand, hinder necessary financial investments and governmental involvement. As mentioned earlier, these legal gaps require a new set of norms. Therefore, the following part will propose a solution towards clear regulations and, consequently, legal certainty.

11.3 No Environment for New Treaty-Making Processes

The space race that followed the beginning of space activities in the 1960s inspired the international community to establish a binding international space law framework with the purpose to safeguard the new emerging activities and to avoid dangers for humanity, e.g. loss of life or property, environmental damages and contamination of outer space, with a set of laws and rules. The failure to adopt new binding treaties at the UN level in the following 30 years indicates that the challenges of the new space missions cannot be overcome by treaty-making in the near future. The discrepancies in political perspectives and the specificities of the consensus decision-making process within the UN COPUOS render it exceptionally difficult for states to agree on new binding provisions. Furthermore, with a growing number of participating states, this situation becomes intensified. It is interesting to note that during the era of the Cold War, the environment of mistrust contributed to the development of legal regimes governing areas not yet conquered. Fear for potential military actions and hope for global peace enabled states, at that time, to reach common grounds for the purposes of their own security. However, nowadays, diverse views and lack of trust among states do not provide for a treaty-making environment.

Even though the Cold War paranoia no longer exists, we have entered another type of real and rigorous competition.[74] Private companies are racing each other in order to secure governmental contracts and be part of national budget funds. One of the reasons why the tendency to develop new technology has shifted towards the private sector is because this is a way to overcome bureaucratic logjams and avoid long delays and cost increases. Nowadays, that speedy action and cost efficiency are essential to achieve new breakthroughs and stay competitive in the market. It is thus important to rely on private actors as the driving engine; only then, national

[73] J. Hurtak, M. Egan, "A MULTILATERAL AGREEMENT FOR A MANNED MISSION TO MARS" p 1–2 (2006); WIPO, International Bureau, Intellectual Property and Space Activities, at http://www.wipo.int/export/sites/www/patent-law/en/developments/pdf/ip_space.pdf.

[74] SpaceNews, OpEd – Competition: Driving Force for the Second Space Age (2004), at http://spacenews.com/oped-competition-driving-force-second-space-age/#sthash.ZNfMWlx1.dpuf.

agencies will be able to progress. This new space age is no longer about the Cold War's heated race to be the first on a new celestial body. Rather, it is a race to secure a place as the leader in science, exploration and technological development. "It's about moving increasingly routine aspects of spaceflight, both manned and unmanned, to the private sector so that government resources can be maximized in bold new missions for space exploration".[75] Therefore, this second space race is about building innovative, competitive, sustainable and inspirational space enterprises.

As a result, new legal regulations to provide clear answers and address the loose interpretations of the articles of the OST cannot be created through amendments to the OST or through a new treaty. Instead, they need to be more flexible, progressive and adaptive to technological advancement. Therefore, alternative legal mechanisms must be explored.

11.4 The ISS Model[76]

A possible model for an alternative legal instrument for outer space could be the framework governing the establishment and operation of the International Space Station (ISS) as it is a model that has proved successful in regulating the most complex space activity so far.

The ISS legal framework is governed by a three-tier structure. At the first level lies the International Space Station Intergovernmental Agreement (IGA),[77] which functions as an overarching umbrella for all legal issues concerning the ISS. At the second level exist four Memoranda of Understanding (MoUs): between NASA and each cooperating space agencies (ESA, CSA, Roscosmos, JAXA).[78] These MoUs

[75] SpaceNews, OpEd – Competition: Driving Force for the Second Space Age (2004), at http://spacenews.com/oped-competition-driving-force-second-space-age/#sthash.ZNfMWlx1.dpuf.

[76] This proposed solution was also elaborated in the forthcoming chapter: "Intellectual Property Rights: an IGA-Based Approach" by A. Nakarada Pecujlic and S. Germann, in the Book "Outer Space: Legal Policy and Practice", Global Law and Business, 2017.

[77] Agreement among the Government of Canada, Governments of Member States of the European Space Agency, the Government of Japan, the Government of the Russian Federation, and the Government of the United States of America concerning Cooperation on the Civil International Space Station, Washington, 29 January 1998, entered into force 27 March 2001 (1998 IGA-Agreement).

[78] Memorandum of Understanding between the National Aeronautics and Space Administration of the United States of America and the Government of Japan concerning Cooperation on the Civil International Space Station, done at Washington, DC, Feb 24, 1998, at http://web.archive.org/web/20091029094939/http://www.nasa.gov/mission_pages/station/structure/elements/nasa_japan.html (MOU NASA/JAXA)

Memorandum of Understanding between the National Aeronautics and Space Administration of the United States of America and the Canadian Space Agency concerning Cooperation on the Civil International Space Station, done at Washington, DC, Jan 29, 1998, at http://web.archive.org/web/20091029094938/http://www.nasa.gov/mission_pages/station/structure/elements/nasa_csa.html (MOU NASA/CSA)

address more practical details of the development of the ISS. At the third level, diverse Implementing Arrangements between the space agencies are featured for the implementation of the MoUs.[79]

The IGA provisions contain very specific and detailed legal norms. For example, Article II para 2 lit c prohibits any basis for asserting a claim to national appropriation over outer space or over any portion of outer space.[80] Other provisions address jurisdiction and control issues, introduce an ISS Crew Code of Conduct and cover exchange of information, goods and intellectual property.[81]

The provisions on cross-waivers of liability as regulated under Article XVI constitute a vivid example of how the involved states achieved agreement on a rule that goes beyond the scope of the space treaties in order to facilitate their activities on the ISS. The importance of this provision is emphasized by the very wording of the article:

> The objective of this Article is to establish a cross-waiver of liability by the partner states and related entities in the interest of encouraging participation in the exploration, exploitation, and use of outer space through the Space Station. This cross-waiver of liability shall be broadly construed to achieve this objective.[82]

Private companies, in their capacities as contractors and subcontractors, are considered to be related entities of Partner States and as such are covered by the cross-waiver clause. As a result, all direct and indirect parties are encouraged to participate

---

Memorandum of Understanding between the National Aeronautics and Space Administration of the United States of America and the Russian Space Agency concerning Cooperation on the Civil International Space Station, done at Washington, DC, Jan 29, 1998, at http://www.nasa.gov/mission_pages/station/structure/elements/nasa_rsa.html (MOU NASA/ROSCOSMOS)

Memorandum of Understanding between the National Aeronautics and Space Administration of the United States of America and the European Space Agency concerning cooperation on the Civil International Space Station, done at Washington, DC, Jan 29, 1998, at http://www.nasa.gov/mission_pages/station/structure/elements/nasa_esa_prt.htm (MOU NASA/ESA).

[79] ESA, International Space Station Legal Framework, at http://www.esa.int/Our_Activities/Human_Spaceflight/International_Space_Station/International_Space_Station_legal_framework.

[80] Article 2 IGA states:

1. The Space Station shall be developed, operated, and utilized in accordance with international law, including the Outer Space Treaty, the Rescue Agreement, the Liability Convention, and the Registration Convention.
2. Nothing in this Agreement shall be interpreted as:
   (a) modifying the rights and obligations of the Partner States found in the treaties listed in paragraph 1 above, either toward each other or toward other States, except *as* otherwise provided in Article 16;
   (b) affecting the rights and obligations of the Partner States when exploring or using outer space, whether individually or in cooperation with other States, in activities unrelated to the Space Station; or
   (c) constituting a basis for asserting a claim to national appropriation over outer space or over any portion of outer space.

[81] Arts 3–21, IGA.

[82] Art 16, IGA.

in space exploration, which is feasible, thanks to the necessary legal certainty provided by the liability cross-waivers.[83]

One more innovation of the IGA can be found in the provisions on technology transfer and IP protection.[84] In this context, the IGA model provides for specific "marking procedures", which protects the property of data and goods.[85] According

---

[83] J. Hurtak, Matthew Jude Egan, "A MULTILATERAL AGREEMENT FOR A MANNED MISSION TO MARS", p 4–5 (2006).

[84] Arts 19–21, IGA.

[85] Article 19 paras 3–4 IGA state: "The Partners agree that transfers of technical data and goods under this Agreement shall be subject to the restrictions set forth in this paragraph. The transfer of technical data for the purposes of discharging the Partners' responsibilities with regard to interface, integration and safety shall normally be made without the restrictions set forth in this paragraph. If detailed design, manufacturing, and processing data and associated software is necessary for interface, integration or safety purposes, the transfer shall be made in accordance with paragraph 1 above, but the data and associated software may be appropriately marked *as* set out below. Technical data and goods not covered by the restrictions set forth in this paragraph shall be transferred without restriction, except as otherwise restricted by national laws or regulations.

(a) The furnishing Cooperating Agency shall mark with a notice, or otherwise specifically identify, the technical data or goods that are to be protected for export control purposes. Such a notice or identification shall indicate any specific conditions regarding how such technical data or goods may be used by the receiving Cooperating Agency and its contractors and subcontractors, including (1) that such technical data or goods shall be used only for the purposes of fulfilling the receiving Cooperating Agency's responsibilities under this Agreement and the relevant MOUs, and (2) that such technical data or goods shall not be used by persons or entities other than the receiving Cooperating Agency, its contractors or subcontractors, or for any other purposes, without the prior written permission of the furnishing Partner State, acting through its Cooperating Agency.

(b) The furnishing Cooperating Agency shall mark with a notice the technical data that are to be protected for proprietary rights purposes. Such notice shall indicate any specific conditions regarding how such technical data may be used by the receiving Cooperating Agency and its contractors and subcontractors, including (1) that such technical data shall be used, duplicated, or disclosed only for the purposes of fulfilling the receiving Cooperating Agency's responsibilities under this Agreement and the relevant MOUs, and (2) that such technical data shall not be used by persons or entities other than the receiving Cooperating Agency, its contractors or subcontractors, or for any other purposes, without the prior written permission of the furnishing Partner State, acting through its Cooperating Agency.

(c) In the event that any technical data or goods transferred under this Agreement are classified, the furnishing Cooperating Agency shall mark with a notice, or otherwise specifically identify, such technical data or goods. The requested Partner State may require that any such transfer shall be pursuant to a security of information agreement or arrangement which sets forth the conditions for transferring and protecting such technical data or goods. A transfer need not be conducted if the receiving Partner State does not provide for the protection of the secrecy of patent applications containing information that is classified or otherwise held in secrecy for national security purposes. No classified technical data or goods shall be transferred under this Agreement unless both parties agree to the transfer.

4. Each Partner State shall take all necessary steps to ensure that technical data or goods received by it under subparagraphs 3(a), 3(b), or 3(c) above shall be treated by the receiving Partner State, its Cooperating Agency, and other persons and entities (including contractors and subcontractors) to which the technical data or goods are subsequently transferred in accordance with the terms of the notice or identification. Each Partner State and Cooperating Agency shall

to these procedures, each of the five space agencies and their contractors and subcontractors mark their technical data and goods with a notice, which indicates the specific conditions of use by the other agencies.[86] This offers the necessary protection (by noting the conditions for the utilization by other users) and provides a necessary environment for smooth and speedy implementation.[87] In that manner, although states retain jurisdiction and control over objects that they launch into outer space, various ISS partners have the right to use each other's research elements and to barter or sell elements of their respective allocations to each other or to third parties.[88]

As noted earlier, the ISS is, politically, the most complex space activity ever undertaken. Yet, collaboration among the five Partner States, the USA, Russia, Canada, Japan and another 11 European states, all represented through their national space agencies (NASA, JAXA, CSA, Roscosmos and ESA, respectively) enabled the ISS vision to materialize.[89] The magnitude of this mission is due to the nature of this space programme as it "brings together international flight crews, multiple launch vehicles, globally distributed launch, operations, training, engineering, and development facilities, communications networks, and the international scientific research community".[90] Each participating state's contributions comprise their expertise in the different ISS segments, and astronauts work collectively to conduct experiments in space.[91] This endeavour has been successful, although the five participating partners originate from diverse national, political and economic backgrounds with different national licensing regimes. Still, they succeeded to commit themselves to the development and operation of this space exploration programme. Therefore, it can be said that the IGA three-tier structure was successfully implemented to clarify legal uncertainties, to provide the necessary certainty for states to engage in this mission and, to a further extent, to encourage their respective private sectors to develop the required technology.

---

take all reasonably necessary steps, including ensuring appropriate contractual conditions in their contracts and subcontracts, to prevent unauthorized use, disclosure, or retransfer of, or unauthorized access to, such technical data or goods. In the case of technical data or goods received under subparagraph 3(c) above, the receiving Partner State or Cooperating Agency shall accord such technical data or goods a level of protection at least equivalent to the level of protection accorded by the furnishing Partner State or Cooperating Agency".

[86] Art 19, IGA.

[87] The Legal Framework for the International Space Station, http://www.unoosa.org/pdf/pres/lsc2013/tech-05E.pdf.

[88] Art VIII, Treaty on Principles Governing the Activities of States in the Exploration and Use of Outer Space, including the Moon and Other Celestial Bodies, 610 UNTS 205, entry into force Oct 10, 1967.

[89] F. Claasen, P. Weber, H. Ripken, V. Sobick, "Promotion of Industrial ISS Utilisation by the German Space Agency", in International Space Station, the next Space Marketplace (eds. G. Haskell, M. Rycroft), p 156 (2000).

[90] International Cooperation, http://www.nasa.gov/mission_pages/station/cooperation/index.html#.VPNQC3zF-8o.

[91] International Cooperation, http://www.nasa.gov/mission_pages/station/cooperation/index.html#.VPNQC3zF-8o.

Several years of negotiation among the partners, including a group of legal experts, were required to accomplish such an agreement.[92] The success of this arrangement emerged from the fact that it addresses one specific space mission, for which solutions have been elaborated in detail and precision. Because specific parameters and requirements (e.g. legal, administrative and programmatic) were defined at an early stage of the negotiations,[93] such a multilateral agreement could only work for individualized missions. Furthermore, it is important to emphasize that general international law and the five space treaties are applicable to the ISS and the activities on board. An agreement like the IGA could only be applied in addition to the existing general framework, as it is a special legal framework tailored to satisfy the needs of a specific mission.

One more advantage of this type of legal mechanisms is that it does not necessarily require amendments to national laws or a ratification process, as states have the possibility to either ratify the agreement or implement such cooperation as executive agreements.[94] Flexibility on the ratification process, on one hand, and hard law obligations,[95] on the other, have rendered the ISS cooperation agreement such a unique and successful framework. The three-tier structure used here could be an example of possible future developments in international space law.[96] In the absence of a comprehensive international treaty, specifically designed legal obligations applicable only to the states involved in specific projects could be the answer.

The solution that this paper suggests does not resolve clashes of principles, such as the non-appropriation principle and the idea of a future civilization and private property on Mars. Nevertheless, it provides a possible model for a legal mechanism among interested partners in regard to issues such as resource utilization,

---

[92] M. Ferrazzani, A. Farand, "A European Perspective on Lessons Learned from the Intergovernmental Agreement on International Space Station Cooperation", 65th International Astronautical Congress, Toronto, Canada (2014).

[93] M. Ferrazzani, A. Farand, "A European Perspective on Lessons Learned from the Intergovernmental Agreement on International Space Station Cooperation", 65th International Astronautical Congress, Toronto, Canada (2014).

[94] The US side decided that the IGA is formally to be treated as an "Executive Agreement", which under US practice does not require ratification by the Senate, as it does not require changes to US laws. See A. Farand, The Space Station Cooperation Framework, ESA Bulletin 94 (1998), at http://www.esa.int/esapub/bulletin/bullet94/FARAND.pdf.

[95] There are some scholars who follow the view that the IGA does not qualify as a "full-fledged treaty in the formal sense of international law", as the USA implemented the IGA as an executive agreement, hence without advice and consent of the US Senate. This debate is beyond the scope of this article; however, following this view, these executive agreements are politically binding and provide for political security and are in practice not less effective; see R. Loosch, The International Space Station – the Legal Framework, p 56, in Les Stations Spatiales Habitées – Aspects Juridiques. Manned Space Stations – Legal Issues. Proceedings of the Colloquium held 7–8 November 1989 in Paris. Edited by Duc Guyenne. ESA SP-305. European Space Agency (1990), p 55.

[96] See further F. Von Der Dunk, "Space Law in the Age of the International Space Station" (2009), https://digitalcommons.unl.edu/cgi/viewcontent.cgi?referer=https://www.google.de/&httpsredir=1&article=1005&context=spacelaw, last accessed 18 January 2018.

liability and technology transfer. As is the case with the ISS, a potential Mars agreement would be in conformity with the existing legal regimes and would add detailed regulations tailored to the specific mission. Such detailed regulation could be more easily negotiated as it would concern only the involved states and the specific mission at hand. A three-tier structure could also be used in the following manner: as an umbrella agreement setting out the general legal norms, with MoUs for more practical regulatory details and, finally, with Implementing Arrangements, which would implement the MoUs with respect to the specificities of each national licensing regime. Through a three-tier approach to a possible Mars agreement, problematic legal issues in the context of the Mars Missions of the near future could successfully be resolved.

11.5 Conclusion

Manned missions to Mars are within our reach and they are likely to happen in our lifetime. In order to achieve optimal results for current and future generations, the legal community will play a central role in materializing these endeavours. According to scholars, "the need to privatize, commercialize and promote outer space activities, operating in and out of the Martian planetary environment, needs to be addressed not only technologically, but also legally".[97] It is, thus, necessary to create a new legal regime or other innovative legal solutions to govern the conduct of the first humans on Mars and to clearly determine the allowed scope of exploration, appropriation and liability.

By proposing to use the ISS legal mechanism as such a solution, this paper intends to emphasize the importance of creating a legal framework that embodies the principles of free and open use of the outer space environment and its planetary bodies on one hand and of structured, specific and precise regulations to foster private investments and provide legal certainty for all participants on the other. Exploring Mars – the most accessible of Earth's planetary neighbours – will require multilateral agreements at all the levels of the expedition. Therefore, the successful legal mechanism of the ISS constitutes an example to be taken into account, as the IGA addresses all the key issues involved, such as technology transfer, non-appropriation, liability, ownership, funding and criminal jurisdiction, and offers a customized set of rules to the requirements of a specific endeavour. Lastly, it utilizes established UN space law principles and fosters greater cooperation between national agencies and the private sector. For a groundbreaking mission, such as a manned mission to Mars, this is exactly what is needed.

[97] J. Hurtak, M. Egan, "A MULTILATERAL AGREEMENT FOR A MANNED MISSION TO MARS" p 8, (2006).

liability and technology transfer. As is the case with the ISS, a potential Mars agreement would be in conformity with the existing legal regimes and would add detailed regulations tailored to the specific mission. Such detailed regulation could be more easily negotiated as it would concern only the involved states and the specific mission at hand. A three-tier structure could also be used in the following manner: an umbrella agreement setting out the general legal norms, with MoUs for more practical regulatory details and finally, with Implementing Arrangements, which would implement the MoUs with respect to the specificities of each national [illegible]

11.5 Conclusion

[illegible]

Chapter 12
The Consolidation of the Five UN Space Treaties into One Comprehensive and Modernized Law of Outer Space Convention: Toward a Global Space Organization

Francesco Gaspari and Alessandra Oliva

12.1 Introduction

The development of the so-called NewSpace industry, together with the emergence of new space players, both governmental and nongovernmental – which did not exist when the current international space law was formed – implies a new approach toward space-related matters including new regulatory considerations. Some of the reasons for such new approach would include the necessity for international regulatory certainty as it is required by investments, the promotion of global public interest in outer space, and the preservation of international peace and security.

The process of consolidation of the current five UN space treaties that govern the uses of outer space constitutes one of the main priorities in the international space legal system. Several proposals have already been put forward envisaging one comprehensive and modernized Law of Outer Space Convention incorporating detailed provisions governing exploration, the use and exploitation of outer space, as well as a specific mechanism for the settlement of disputes.

Before entering *in medias res* and investigating whether the current international space law provides an adequate framework to address the new challenges of human space activities, it seems necessary to present a general overview of the state of the art of the current regulatory system.

This paper has been developed and written with the joint efforts of both the authors. However, paragraphs 1, 2, 3, and 7 can be attributed to Prof. Gaspari while paragraphs 4, 5, and 6 to Mrs. Oliva.

F. Gaspari (✉)
G. Marconi University, Rome, Italy
e-mail: f.gaspari@unimarconi.it

A. Oliva
Italian Ministry of Foreign Affairs and International Cooperation, Rome, Italy

G. D. Kyriakopoulos, M. Manoli (eds.), *The Space Treaties at Crossroads*,
https://doi.org/10.1007/978-3-030-01479-7_12

12.2 The Current Space Law Regulatory System: An Overview

Space law is described as the body of law governing space-related activities. Space law, like general international law, comprises a variety of international agreements, treaties, conventions, and United Nations General Assembly resolutions as well as rules and regulations of international organizations. Space law is most often associated with the rules, principles, and standards of international law that appear in the five international treaties and five sets of principles governing outer space, all developed under the auspices of the United Nations.[1]

The international space law system also includes space law *fora*, namely, the United Nations Office for Outer Space Affairs (UNOOSA),[2] the Legal Subcommittee (LSC) of the United Nations Committee on the Peaceful Uses of Outer Space (COPUOS), which is the primary international forum for the development of laws and principles governing outer space and its resources,[3] and the International Institute of Space Law (IISL).[4]

By ratifying the Outer Space Treaty of 1967, 98 nations agreed that the use of outer space would be the "province of mankind," that all nations would have the freedom to "use" and "explore" outer space, and that both these provisions must be adhered to in a way to "benefit all mankind."[5]

---

[1] See *Space Law*, at http://www.unoosa.org/oosa/en/ourwork/spacelaw/index.html. See also S. Marchisio, *Lezioni di diritto aerospaziale*, Rome, 1993, p. 11 ff. In general, on the role of the UN in the development of space law see V. Kopal, *The Role of the United Nations Declarations of Principles in the Progressive Development of Space Law*, in *Journal of Space Law*, 1988, p. 5–20; Id., Evolution of the main principles of space law in the institutional frame work of the United Nations, in *Journal of Space Law*, Vol. 12, 1984, p. 12–25; N. Jasentuliyana, *Strengthening International Space Law: the Role of the United Nations*, in *International Organisations and Space Law: Their Role and Contributions*, in *Proceedings of the Third ECSL Colloquium, (Perugia, 1999)*, Noordwijk, 1999, p. 87–95; Id., *Treaty Law and Outer Space: Can the United Nations play an effective role?* in *Annals of Air and Space Law*, Vol. 11, 1986, p. 219–227.

[2] The UNOOSA provides information and advice to governmental bodies, nongovernmental organizations, as well as the general public. This office also provides copies of the treaties, declarations, working papers, minutes, and other documents related to the creation and development of space, and it prepares legal studies and informative documents on space law and acts as the secretariat for the Legal Subcommittee of the United Nations Committee on the Peaceful Uses of Outer Space (COPUOS).

[3] Member states use this forum to come together, discuss, deliberate, and make decisions concerning space law. On COPUOS see, inter alia, S. Marchisio, *Il ruolo del Comitato delle Nazioni Unite sugli usi pacifici dello spazio extraatmosferico (COPUOS)*, in *Scritti in onore di Giorgio Badiali*, Rome, 2007, p. 221–236. This author points out as currently one of the main tasks of COPUOS is to broaden the universal acceptance of the five treaties, calling upon states to consider the benefits that can derive from that (therein, p. 12).

[4] The IISL cooperates with international organizations and national institutions in the field of space law and helps to foster space law development.

[5] See Preamble and Article 1.

The province of mankind principle was never specifically defined, thus allowing the superpowers to pursue their interests.[6] As a result, the basic foundation of international space law consists of five international space treaties, along with various written resolutions and declarations.[7]

The main international space treaty is the Treaty on Principles Governing the Activities of States in the Exploration and Use of Outer Space, including the Moon and Other Celestial Bodies (Outer Space Treaty),[8] generally considered as the "constitution" for the uses of outer space.[9] It represents the first effort to regulate activities taking place in outer space, as it established several principles of international law and incorporated the principles of peaceful use of outer space, the cooperation between space-faring nations, and the extension of the rule of law into outer space. This is why it is considered the cornerstone of international space law, and the principles set out therein are considered the source and substance from which all the subsequent four agreements were derived.[10]

---

[6] See E. Weeks & E. Dahlstrom (Eds), *Space Mining and Use of Space Natural Resources (Chapter 16)*, in R.S. Jakhu & J.N. Pelton (Coordinated by), *Global Space Governance: An International Study*, IASL-McGill University, Springer, Cham, 2017, p. 379–416; N. Jasentuliyana, *Space Law: Development and Scope*, Westport, Connecticut, 1992; N. Jasentuliyana and R.S.K. Lee (Eds), *Manual on Space Law* (4 volumes), Dobbs Ferry, New York, 1979–1981.

[7] The five international treaties are as follows: The Treaty on Principles Governing the Activities of States in the Exploration and Use of Outer Space, including the Moon and Other Celestial Bodies (the "Outer Space Treaty") – adopted on 19 December 1966 in General Assembly resolution 2222 (XXI), opened for signature on 27 January 1967, entered into force on 10 October 1967; the Agreement on the Rescue of Astronauts, the Return of Astronauts, and the Return of Objects Launched into Outer Space (the "Rescue Agreement") – adopted on 19 December 1967 in General Assembly resolution 2345 (XXII), opened for signature on 22 April 1968, entered into force on 2 December 1968; the Convention on International Liability for Damage Caused by Space Objects (the "Liability Convention") – adopted on 29 November 1971 in General Assembly resolution 2777 (XXVI), opened for signature on 29 March 1972, entered into force on 1 September 1972; the Convention on Registration of Objects Launched into Outer Space (the "Registration Convention") – adopted on 12 November 1974 in General Assembly resolution 3235 (XXIX), opened for signature on 14 January 1975, entered into force on 15 September 1976; and the Agreement Governing the Activities of States on the Moon and Other Celestial Bodies (the "Moon Agreement") – adopted on 5 December 1979 in General Assembly resolution 38/68, opened for signature on 18 December 1979, entered into force on 11 July 1984. Each treaty is available online at http://www.unoosa.org/pdf/publications/STSPACE11E.pdf.

[8] *Treaty on Principles Governing the Activities of States in the Exploration and Use of Outer Space, Including the Moon and Other Celestial Bodies*, 27 January 1967, 610 UNTS 205, 18 UST 2410, TIAS No 6347, 6 ILM 386 (entered into force on 10 October 1967) [Outer Space Treaty].

[9] See R. Jakhu, *Legal Issues Relating to the Global Public Interest in Outer Space*, 2005, p. 3, available online http://www.cissm.umd.edu/publications/legal-issues-relating-global-public-interest-outer-space-0; E. Weeks & E. Dahlstrom (Eds), *Mining and Utilization of Space Natural Resources (Chapter 15)*, cit. The Outer Space Treaty was ratified by 96 nations and signed by another 27 states. See UN Committee on the Peaceful Uses of Outer Space, *Report of the Legal Subcommittee on Its Fortieth Session*, United Nations' Document A/AC.105/763, 24 April 2001, available at the webpage http:www.oosa.unvienna.org/Reports/AC105_763E.pdf.

[10] R. Berkley, *Space Law* Versus *Space Utilization: The Inhibition of Private Industry in Outer Space*, in *Wisconsin International Law Journal*, Vol. 15, 1997, p. 421; E. Weeks & E. Dahlstrom (Eds), *Mining and Utilization of Space Natural Resources (Chapter 15)*, cit.

By contrast, the Agreement governing the Activities of States on the Moon and Other Celestial Bodies (Moon Agreement)[11] is often treated as though it were not part of the body of the international space law. Equally, extensive debates have taken place on whether or not the Moon Agreement comprises a valid part of the international law.[12] The Moon Agreement entered into force in 1984, based on a five-state ratification consensus procedure, agreed upon by the members of the COPUOS. Still today, very few nations have signed and/or ratified it.[13]

The other three outer space treaties experienced a high level of international cooperation in terms of signature and ratification. However, the Moon Treaty moved further than them, by introducing the common heritage of mankind concept and by imposing specific obligations on the parties engaged in the exploration and/or exploitation of the outer space.[14] The Moon Treaty explicitly designates the Moon and its natural resources as part of the common heritage of mankind.[15]

As it has been pointed out, among the key principles contained in the five international space treaties is the understanding that the use of outer space is deemed to be the province of mankind, as distinct from the common heritage of mankind.[16]

The drafters of the Outer Space Treaty initially focused on solidifying broad terms with the intent to create more specific legal provisions later.[17] Given that

[11] *Agreement governing the Activities of States on the Moon and Other Celestial Bodies*, 5 December 1979, 1363 UNTS 3 (entered into force 11 July 1984) [Moon Agreement].

[12] E. Weeks & E. Dahlstrom (Eds), *Mining and Utilization of Space Natural Resources (Chapter 15)*, cit.

[13] As at 1 January 2015, only 16 states (Australia, Austria, Belgium, Chile, Kazakhstan, Kuwait, Lebanon, Mexico, Morocco, the Netherlands, Pakistan, Peru, the Philippines, Saudi Arabia, Turkey, Uruguay) have ratified it, and 4 states (France, Guatemala, India, and Romania) in addition have signed but not ratified

[14] E. Weeks & E. Dahlstrom (Eds), *Mining and Utilization of Space Natural Resources (Chapter 15)*, cit.

[15] See S. Marchisio, *Il ruolo del Comitato delle Nazioni Unite*, cit., p. 12. On the *Common Heritage of Mankind* concept, see also K. Baslar, *The Concept of the Common Heritage of Mankind in International Law*, the Hague-Boston, London, 1998; S. Gorove, *The Concept of "Common Heritage of Mankind": A Political, Moral or Legal Innovation?*, in *San Diego L. Rev.*, Vol. 9, 1972, p. 390; G.M. Danilenko, *The Concept of the Common Heritage of Mankind in International Law*, in *Annals Air & Space* Law, Vol. 13, 1988, p. 247; R. Wolfrum, *The Principle of the Common Heritage of Mankind*, in *Zeitschrift für Auslandisches Offentliches Recht und Volkerrecht*, Vol. 43, 1983, p. 312; K. Tatsuzawa, *Political and Legal Meaning of the Common Heritage of Mankind*, in *Proceeding of the twenty-ninth Colloquium on the Law of Outer Space, Am. Inst. of Aeronautics & Astronautics ed.*, 1986; C.Q. Christol, *The Common Heritage of Mankind Provisions in the 1979 Agreement Governing the Activities of States on the Moon and Other Celestial Bodies*, in *International Law*, Vol. 14, 1980, p. 429; V. Kopal, *Outer Space as a Global Common*, in *Proceedings of the* fortieth *Colloquium on the Law of Outer Space, Am. Inst. of Aeronautics & Astronautics ed.*, 1997.

[16] E. Weeks, *Outer Space Development, International Relations, and Space Law: A Method for Elucidating Seeds*, Newcastle upon Tyne, 2012, p. 44; E. Weeks & E. Dahlstrom (Eds), *Mining and Utilization of Space Natural Resources (Chapter 15)*, cit.

[17] This is the reason why the members of the COPUOS later elaborated on the Outer Space Treaty norms through the adoption of the "three supplemental agreements" – The Rescue and Return

important issues such as the environment, public health, and distribution of benefits were left open, key actors were involved in space law negotiations which led to legal norms later embodied in the Moon Agreement.[18]

12.3 The "NewSpace Industry" and the Need for a New Regulatory System

Since the 1950s – when the first steps of the modern space adventure of humans took place[19]– the relationship between human being and outer space is continuously expanding.

During the first era, from the launch of Sputnik I in 1957 to the beginning of the 1970s, also defined as the "scientific research phase," the development of space regime was dominated by two superpowers (the United States and USSR) – together with a group of other states operating through the United Nations[20] – and was characterized by the establishment of new capabilities and new systems aiming to demonstrate their scientific and technologic superiority and, through it, the supremacy of their respective political and economic systems.[21] Later on, within the so-called operating phase (from the 1970s to the 1980s) and with the establishment of the first international organizations, especially in the telecommunications sector, the goal was, on the one hand, the management of specific satellite applications and, on the other hand, the response to commercial considerations and to economic needs that had just appeared.[22] In the current era, which begun in the 1980s and is also defined as *exploiting phase*, the objective of space activities seems to be the search for economic profit.[23]

This new era in space activities begun with the 30-year US Space Shuttle Program, which, in 1981, proposed a new space mission approach: "humanity's first

Agreement of 1968, the Liability Convention of 1973, and the Registration Convention of 1976: see N.L. Griffin, *The Americans and the Moon Treaty*, in *Journal of Air Law and Commerce*, Vol. 46, 1981, p. 730, especially p. 733–734.

[18] A. Bini, *The Moon Agreement in the 21st century*, in *Acta Astronautica*, Vol. 67, Issues 3–4, August–September 2010, p. 496–501.

[19] We especially refer to the launch into orbit of the first Sputnik in 1957.

[20] See *Outer Space Development*, at https://sites.google.com/site/outerspacedevelopment/history-of-spaceflight.

[21] See M. Couston, *Droit spatial économique. Régime applicables à l'éxploitation de l'espace*, SIDES, 1994, p. XII–XIII.

[22] In this respect, the experience of Arianespace SA is an important example. Arianespace SA is a European multinational company founded in the 1980s as the world's first commercial launch service provider. For further information, see http://www.arianespace.com/about-us/.

[23] See M. Couston, *Droit spatial économique*, cit., p. XIII.

reusable spacecraft."[24] In fact, the Space Shuttle was not only a vehicle that could be used for one sole mission, but it was also able to remain in orbit and to be back toward Earth like an aircraft, enabling transports of freight between Earth and space and contributing, inter alia, to the construction of the International Space Station.[25] With the Space Shuttle, outer space begun to be at the disposal of humanity also under a commercial point of view.[26] As a consequence, the third phase is characterized, above all, by a process of privatization and by the increased financial participation of the private sector in space programs, thus changing the nature of actors involved. Such development might thus enable states to reduce the use of public resources and to share the weight of significant space investment allocation.

Therefore, the scenario has dramatically changed, and it still evolves today in the light of technological developments.[27] In the future, the main thrust of space activities will be realized in the context of what has already started to be called as "NewSpace industry," in which new and nontraditional space players take part.

The scope of the "NewSpace industry" and the emergence of new players (together with the fast-growing technological developments and the need for private investments) call for a new international regulatory framework of space activities to cover the deficiencies of current and often obsolete rules.

---

[24] See Space Shuttle Era at http://www.nasa.gov/mission_pages/shuttle/flyout/index.html.

Between the first launch on April 12, 1981, and the final landing on July 21, 2011, NASA's Space Shuttle fleet flew 135 missions. NASA's Space Shuttle fleet began setting records with its first launch on April 12, 1981, and continued to set high marks of achievement and endurance through 30 years of missions. Starting with Columbia and continuing with Challenger, Discovery, Atlantis, and Endeavor, the spacecraft has carried people into orbit repeatedly, launched, recovered and repaired satellites, conducted cutting-edge research, and built the largest structure in space, the International Space Station. The final Space Shuttle mission, STS-135, ended on July 21, 2011, when Atlantis rolled to a stop at its home port, NASA's Kennedy Space Center in Florida: ibidem.

[25] See S. Marchisio, *Lezioni di diritto aerospaziale*, cit., p. 13.

[26] See S. Hobe, *Current and Future Development of International Space Law*, in *Proceedings United Nations/Brazil Workshop on Space Law "Disseminating and Developing International and National Space Law: the Latin America and Caribbean Perspective*", 2005, p. 14.

[27] Most notably and importantly, space activities provide practical benefits to all states. Thirteen states have independent launch capability; over 60 countries operate their satellites; there are over 70 space agencies; currently more than 1200 satellites are being operated; thousands of new satellites (especially small sats) are planned to be launched in the near future; the global space industry has become a $320 billion economic activity annually; the private sector now is spearheading unprecedented movements in space; there are serious emerging challenges to the sustainability of space activities of all nations; and the international space law-making process is getting stagnated fast. See the presentation of the International Conference on New Challenges in Space Law, *The Space Treaties at Crossroads: Considerations for* de lege ferenda, jointly organized by the Department of International and European Studies Faculty of Law National and Kapodistrian University of Athens, Greece, and by the Institute of Air and Space Law and Centre for Research in Air and Space Law Faculty of Law McGill University, Montreal, Canada, and held in Athens, Greece, on 28 and 29 August 2015. The document is available in the conference's website at http://www.nb.org/%CE%9A%CE%95%CE%99%CE%9C%CE%95%CE%9D%CE%91/International_Conference_on_New_Challenges_in_Space_Law.

The five space treaties were drafted and adopted within the United Nations umbrella between the 1960s and 1970s and lay down a series of norms and principles that remain valid still today. However, given the era in which these principles were developed, they did not explicitly address neither the wide involvement of the private sector nor the unprecedented technological developments that have been massively transforming the outer space sector during the last decades.[28]

Such developments require the establishment of a new regulatory approach as they call for new rules to address emerging needs. In other words, new regulation must take into account new scenarios that stem from technological developments and, consequently, the needs of new players. Such novel schemes will ensure legal certainty at the international level, which is necessary to attract investments, to promote global public interest in outer space, and to maintain international peace and security.[29]

As it is well established, legal certainty constitutes an important general principle of international law and of European Union law.[30] It is also one of the most important principles of public international law recognized by the European Court of Justice (ECJ).[31]

---

[28] See R. Jakhu, *Legal Issues*, cit., p. 66. Moreover, see 59th International Astronautical Congress, Plenary 3, 30 September 2008.

[29] As it has been pointed out, "*the exploitation of the natural resources of the Moon and other celestial bodies is a risky and expensive task*", and "*carrying out activities in outer space is a hazardous business*": See, in this respect, B.M. Hoffstadt, *Moving the Heavens: Lunar Mining and the "Common Heritage of Mankind" in the Moon Treaty*, in *UCLA L. Rev.*, Vol. 42, 1994, p. 575, 580 & n. 24.

According to an author, the two main risks deriving from the exploitation of the natural resources of the Moon and other celestial bodies are the possible mistakes and unexpected events that can occur in missions and that can cause irremediable damages (such as the destruction of a space object and the death of its occupants) as well as the huge investments required to develop safe and reliable technology to exploit extraterrestrial resources and to establish, as well as maintain, a permanent manned lunar basis. Therefore, given that "*These two elements create serious obstacles to the actual commencement of the exploitation of lunar resources*" the Author concludes that "*A method to soften their negative impact may be the establishment of a legal regime to regulate exploitation.*": F. Tronchetti, *The Moon Agreement in the 21st Century: Addressing it potential role in the Era of commercial exploitation of the natural resources of the Moon and other celestial bodies*, in *Journal of Space Law*, Vol. 36, p. 509–510.

[30] The legal certainty is recognized as one of the general principles of European Union law by the European Court of Justice since the 1960s: see D. Chalmers, *European Union law: text and materials*, Cambridge, 2006, p. 454.

[31] In the context of the EU legislation, this means that the law must be certain in that it must be clear and precise and its legal implications foreseeable, especially when applied to financial matters. The adoption of laws which will have legal effect in the European Union must have a proper legal basis. Legislation in member states which implements European Union law must be worded so that it is clearly understandable by those who are subject to the law. Moreover, the ECJ has given a concrete scope of application to the principle of legal certainty in order to escape its tautological nature and to clarify its content. As a result, under the principle of legal certainty, several rules have been established by the ECJ: the principle of non-retroactivity of administrative acts, the principle of good faith, the principle of *patere legem quam ipse fecisti*, the principle of vested or acquired rights, and the principle of legitimate expectations. On the principle of legal certainty and on those rules, see, among others, A. Kaczorowsky, *European Union law*, London-New York, 2008, p. 232; G. Tesauro, *Diritto dell'Unione europea*, Padua, 2012, p. 105 ff.

Within international law, legal certainty is recognized as a central characteristic of the rule of law.[32]

Similarly, the need for legal certainty is required, inter alia, in the field of exploitation of natural resources of outer space, as the five UN space treaties are not tailored to address the current needs, thus causing the reluctance of potential investors. As a result, international space law will continue to constitute a hindrance to space enterprise and space industry,[33] will possibly prevent space activities from further development, or will even contribute to a space race with multiple negative implications.

12.4 Attempts Toward a New Legal Framework for Outer Space Activities and Subsequent Issues of Implementation

The emergence of new key players in space activities and the consequent existence of new needs render essential the establishment of a new regulatory framework for outer space. Although space law has shown an extraordinary capacity to adapt itself to new challenges – raised mainly by a continuous technological progress that constantly broadens space boundaries – the need to preserve the peaceful use of outer space requires adjustments. The discussion on a comprehensive framework to govern the uses of outer space is not new. In fact, the Moon Agreement already foresaw the establishment of a specific regime with reference to the activity of exploitation of natural resources. Nevertheless, nowadays, the possibility that states and private space actors implement such commercial programs is real and requires a broadly agreed legal regime. Indeed, the five space treaties were adopted in an era when governments were the main actors in space activities. Although Art. VI of the Outer Space Treaty recognizes a possible role to be played by "nongovernmental entities," the said article can be interpreted from various angles.[34]

---

[32] See D. Maresca, *Regulation of Infrastructure Markets: Legal Cases and Materials on Seaports, Railways and Airports*, Heidelberg, 2013, p. 26.

[33] See T.G. Nelson, *The Moon Agreement and Private Enterprise: Lessons from Investment Law*, in *ILSA J. Int'l & Comp. L.*, Vol. 17, 2011, p. 393–416; L.M. Fountain, *Creating Momentum in Space: Ending the Paralysis Produced by the 'Common Heritage of Mankind' Doctrine*, in *Conn. L. Rev.*, Vol. 35, 2003, p. 1753, especially p. 1759; C. Buxton, *Property in Outer Space: The Common Heritage of Mankind Principle* vs. *the 'First in Time, First in Right' Rule of Property Law*, in *Journal of Air Law & Commerce*, Vol. 69, 2004, p. 689; M.E. Schwind, *Open Stars: An Examination of the United States Push to Privatize International Telecommunications Satellites*, in *Suffolk Transnational Law Review*, Vol. 10, 1986, p. 87, especially p. 93.

[34] See *3rd International ECSL Colloquium on International Organizations and Space Law: Their Role and Contributions*, held in Perugia, Italy, on May 6th and 7th 1999, p. 91.

The issue concerning the exploitation of natural resources in outer space is an example of such a lack of legal certainty. As stated by the International Institute of Space Law (IISL):

> At present, international space legislation does not include detailed provisions with regard to the exploitation of natural resources of outer space, the Moon and other celestial bodies […]. The IISL is of the opinion that a specific legal regime for the exploitation of such resources should be elaborated through the United Nations, on the basis of present international space law, for the purposes of clarity and legal certainty in the near future.[35]

The same problem could be raised for other rules established in the Outer Space Treaty and in the other space treaties, thus leading to a general legal uncertainty with reference to possible new investments in space projects.

Over the years, a number of proposals have been put forward to address the question. Three main ideas have emerged in particular: the ratification of the Moon Agreement, the consolidation of the UN space treaties and principles into one single comprehensive Law of Outer Space Convention, and, closely connected to the latter, the establishment of a Global Space Organization.

With reference to the Moon Agreement, the low number of state parties, none of which are currently leading space-faring nations – mainly due to the presence of the Common Heritage of Mankind Principle[36] – prevents it from being an effective legal instrument. Even though such an agreement might be a potential point of reference or starting point since it affirms at Art. 11.5 that "States parties to this Agreement hereby undertake to establish an international regime […] to govern the exploitation of the natural resources of the Moon as such exploitation is about to become feasible," it is not possible to ignore those political factors that have hindered an increase in the number of ratifications. Therefore, although many international documents and resolutions have called for the accession to the Agreement,[37] it is largely believed that a new and (especially) shared international regime dealing with a wider range of outer space issues needs to be adopted.[38]

---

[35] Statement of the Board of Directors of the International Institute of Space Law (IISL), 22 March 2009, available online at http://www.iislweb.org/docs/Statement%20BoD.pdf.

[36] See R. Jakhu, *Legal Issues*, cit., p. 66–67.

[37] UN Doc. A/AC.105/C.2/L.272, *Joint statement on the benefits of adherence to the Agreement Governing the Activities of States on the Moon and Other Celestial Bodies by States parties to the Agreement*, Vienna March 31 to April 11 2008; ILA Resolution 1/2002 with regard to the Common Heritage of Mankind Principle in the Moon Agreement; Statement of the Board of Directors of the International Institute of Space Law (IISL), cited above.

[38] See R. B. Bilder, *A Legal Regime for the Mining of Helium-3 on the Moon: U.S. Policy Options*, in *Fordham International Law Journal*, Vol. 33, Issue 2, 2009, p. 248; See also P. Martinez and J. White, *The Adequacy of the Current Legal and Regulatory Framework Regarding Commercial Activities in Outer Space – A Developing Country Perspective*, Discussion Paper, Workshop Proceedings – *International and Interdisciplinary Workshop on Policy and Law Relating to Outer Space Resources: Examples of the Moon, Mars, and other celestial bodies*, McGill University, Centre for Research of Air and Space Law, Montreal, 2006, p. 216.

12.5 A Possible Model for a Single Law of Outer Space Convention

A consolidation of the outer space treaties and principles into a single and renewed international convention is an adequate solution.[39] This would be a highly complex task to fulfill due to the fact that the number of states parties to the UN space treaties has progressively decreased, as demonstrated by the difference between the number of ratifications to the Outer Space Treaty (103) and those to the Moon Agreement (only 17).[40] For this reason, in order to consolidate the treaties into a single Law of Outer Space Convention, we believe that two processes should be taken into account: the codification of customary rules with regard to the use of outer space and the progressive development of international space law.[41] A possible model to look at is the 1982 United Nations Convention on the Law of the Sea (UNCLOS).[42] Indeed, not only does it regard the regulation of an area that is placed beyond national jurisdiction, but it also codifies international customary law as well as establishes new international rules to facilitate the further development of the international law of the sea.

UNCLOS could be considered a viable example because, unlike the Outer Space Treaty, it offers a specific regime concerning the exploitation of seabed resources consisting of its Part XI, the III and IV Annexes, the 1994 Implementing Agreement relating to deep seabed mining, and the so-called Mining Code. This latter instrument refers to a "comprehensive set of rules, regulations and procedures issued by the International Seabed Authority to regulate prospecting, exploration and exploitation of marine minerals in the international seabed Area (defined as the seabed and subsoil beyond the limits of national jurisdiction)."[43] The key aspect of such a system is that these rules are closely connected to the general legal framework represented by the UNCLOS, thanks to the activity of one of its bodies, namely, the Seabed Authority.

---

[39] A proposal that according to some states should be considered by the COPUOS Legal Subcommittee. See, for example, the working paper A/AC.105/C.2/L.23 submitted by China, Greece, and the Russian Federation.

[40] COPUOS Legal Subcommittee, Fifty-fourth session (Vienna, 13–24 April 2015), *Status and application of the five United Nations treaties on outer space*, doc. A/AC.105/C.2/2015/CRP.8, available online at http://www.unoosa.org/pdf/limited/c2/AC105_C2_2015_CRP08E.pdf.

[41] The codification and the progressive development of international law can both be found in Art. 13 of the UN Charter, according to which "The General Assembly shall initiate studies and make recommendations for the purpose of [...] encouraging the progressive development of international law and its codification."

[42] The convention lays down a comprehensive regime of law and order in the world's oceans and seas establishing rules governing all uses of the oceans and their resources. It was opened for signature on 10 December 1982 in Montego Bay, Jamaica, and entered into force in accordance with its Article 308 on 16 November 1994, 12 months after the date of deposit of the sixtieth instrument of ratification or accession. For further information on the convention, please see http://www.un.org/depts/los/convention_agreements/convention_overview_convention.htm.

[43] See International Seabed Authority, *The Mining Code*, available online at https://www.isa.org.jm/mining-code.

Another important aspect of the UNCLOS that could be useful for a new outer space regime is that it contemplates a specific regulation for the settlement of disputes not only at a general level (Part XV) but also at a more specific one within Part XI. Indeed, UNCLOS has established an expert body within the Tribunal for the Law of the Sea and the Seabed Disputes Chamber[44]: the major element of this system is that parties to such disputes may be states parties, the International Seabed Authorities, and the Enterprise as well as State enterprises and natural or juridical persons.[45] This is a very important aspect to take into account, especially considering the growing role and the increasing number of private actors in the outer space activities, as it was demonstrated in the first part of this paper. Indeed, investment of large sums of capital into space projects requires foreseeability of the possible legal implications and their consequences on investment and intellectual property rights.[46]

The model represented by the UNCLOS can be considered successful given the high number of states parties,[47] despite the absence of the United States. Such a structure could be a relevant point of reference for the future regime of outer space: a comprehensive agreement attached to which there could be protocols or annexes concerning the exploitation activities as well as specific provisions for the settlement of disputes and for property rights. This kind of framework also provides the necessary flexibility so as to adequately address the continuous technological progress that affects human activities in outer space by adopting new protocols or annexes, thus avoiding the complex procedures that usually characterize the amendment or the revision of a treaty.[48] An example of a flexible international convention is provided by the 1950 European Convention on Human Rights of the Council of Europe (ECHR).[49] In fact, it is considered a "living instrument" because, although it was adopted in the 1950s, it continuously adapts itself to the actual needs of an ever-changing society and to the contemporary living conditions through the adoption of new protocols, besides the key role played by its court.[50]

---

[44] UNCLOS, Annex VI, Section 1, Article 14: "*A Seabed Disputes Chamber shall be established in accordance with the provisions of section 4 of this Annex. Its jurisdiction, powers and functions shall be as provided for in Part XI, section 5.*"

[45] UNCLOS, Part XI, Section 5, Article 187.

[46] See A.V. Yakovenko, *World Space Organization: pro et contra*, in *3rd International ECSL Colloquium on International Organizations and Space Law*, cit., p. 370.

[47] As of April 3rd 2018, there are 168 member states: see http://www.un.org/depts/los/reference_files/chronological_lists_of_ratifications.htm.

[48] See also R. Jakhu, *Legal Issues*, cit., p. 67.

[49] The ECHR was opened for signature in Rome on 4 November 1950 and came into force in 1953. Since its adoption in the 1950s, the convention has been amended a number of times and supplemented with many rights in addition to those set forth in the original text. For further details, please see https://www.echr.coe.int/Pages/home.aspx?p=basictexts&c.

[50] European Court for Human Rights, judgment of June 30, 1993, *Sigurður Sigurjónsson/Iceland*, in *Racc.*, 1993, p. 264

Nevertheless, the difficulties that could arise from the introduction of a Law of Outer Space Convention should also be pointed out. The main problem in adopting this approach is the need for new international negotiations, which not only would require a long period of time, but they would also provide no guarantee that, in the end, states will sign and ratify the final accord, as the Moon Agreement process has shown. In addition, some states, especially space-faring nations, might take the occasion to weaken some of the key provisions of the Outer Space Treaty, including those that create global public interest in outer space,[51] such as Art. I.1. Such potential complications, however, cannot prevent space law from being updated and modernized to be able to tackle challenges. For this reason, the path toward a Law of Outer Space Convention needs to be followed.

The natural forum in which such a convention might be negotiated is the UNCOPUOS. The United Nations is not only the universal organization in which almost all states of the international community have a seat, but, more importantly, it is also the core of international cooperation, which should be the foundation for space activities. Moreover, the UNCOPUOS is the only body of the UN system dealing with the space sector.[52] It has had a major role in shaping and enhancing space law, and it still pays its fundamental contribution by adopting a number of documents aiming to facilitate the implementation of the outer space rules and to clarify their interpretation.[53] This UN body should thus continue to play such a key function not only by introducing the necessary steps toward the Law of Outer Space Convention but also by remaining the political focal point for space players.

12.6 Toward a Global Space Organization

The UN membership is only open to states, thus excluding all other possible entities, such as space agencies or private actors.[54] However, the UN space treaties provide that states and bodies such as international organizations can enjoy the

---

[51] See R. Jakhu, *Legal Issues*, cit., p. 66.

[52] See S. Marchisio, *Il ruolo del Comitato delle Nazioni Unite*, cit., p. 21.

[53] An example is provided by the 3-year work devoted to the clarification of the term *launching State* and the adoption by the UN General Assembly of the resolution *Application of the Concept of the "Launching State"*, G.A. Res. 59/155 (Dec. 10, 2004). See S. Marchisio, *Il ruolo del Comitato delle Nazioni Unite*, cit., p. 18. A more recent initiative of the COPUOS is the *Protocol to the Convention on International Interests in Mobile Equipment on Matters Specific to Space Assets*, adopted in Berlin in March 2012, but not yet in force. It concerns space assets and the protection of private investments in order to facilitate space activities' commercialization and development. The text of the protocol is available online at http://www.unidroit.org/instruments/security-interests/space-protocol.

[54] Art. 4.1 of the Charter of the United Nations: "*Membership in the United Nations is open to all other peace-loving states which accept the obligations contained in the present Charter and, in the judgment of the Organization, are able and willing to carry out these obligations.*"

rights and be bound by the obligations provided for therein.[55] This is why the need for such a convention is critical, and the need for a new international body where interests of all parties involved would be adequately represented. Even though the creation of a new international organization might be seen as a complex process to pursue, it should be pointed out that a similar trend has been followed by the international community with reference to those sectors assuming a growing significance, such as outer space, in order to strengthen the international partnerships. Besides, "cooperation" should be the "keyword" for space projects, and, as it is well known, international organizations are the institutionalization of international cooperation.[56]

These constitute the main reasons why it is deemed necessary to create an international organization specifically devoted to outer space. The element that would mark the difference in comparison with all the other intergovernmental organizations would be that not only the states but also private companies, national and regional space agencies like the European Space Agencies, as well as other international organizations, for instance, the EU, should have a voice. Such an approach would also provide answers to the issue of the growing involvement of the private sector in outer space activities.

The example of INTELSAT as a point of reference is one option. Before it became an international organization, two separate agreements were adopted in order to establish an interim arrangement for a global commercial communications satellite system,[57] owning and managing a constellation of communication satellites providing international broadcast services. In fact, INTELSAT established the first commercial global satellite communications system in 1965, and for the first time, individuals, businesses, and governments could communicate instantly, reliably, and simultaneously. In 1971 its title changed to International Telecommunications Satellite Organization and was finally privatized in 2001 when it was attributed its current name, INTELSAT.[58]

The example of INTELSAT, however, considers only a sector of space activities, albeit it represents one of the most important kinds of activities. If a Law of Outer Space Convention is adopted, it will be vital to establish an international organization addressing the wide range of space activities and interests. In other

---

[55] See, for example, the declaration made by the European Telecommunications Satellite Organization Intergovernmental Organization (EUTELSAT|IGO), the European Space Agency (ESA), and the European Organization for the Exploitation of the Meteorological Satellites (EUMETSAT) accepting the rights and obligations provided for in the Convention on Registration of Objects Launched into Outer Space: please see http://www.unoosa.org/pdf/limited/l/AC105_2011_CRP12E.pdf, especially p. 10.

[56] On states' obligation to cooperate with each other and to promote cooperation in the exploration and use of outer space, including the Moon and other celestial bodies, see R. Jakhu, *Legal Issues*, cit., p. 18, and *passim*.

[57] Agreement Establishing Interim Arrangements for a Global Commercial Communications Satellite System and Special Agreement, done on August 20, 1964. See also S. Marchisio, *Lezioni di diritto aerospaziale*,, Rome, 2000, p. 42.

[58] On INTELSAT history, see http://www.intelsat.com/about-us/our-history/.

words, a global space convention should envisage a Global Space Organization. Such an institution – that could be linked to the United Nations by becoming a new specialized agency – should have as its main goal the concrete creation of the conditions under which all countries would benefit from space technology and science. At the same time, it should ensure adequate reward and legal certainty for those entities investing in space projects. While the UNCOPUOS should remain the main political forum, the new Global Space Organization should be the "operating arm" of international cooperation in outer space and the place where all actors involved could develop new space projects and activities in a clearer and more defined legal framework.

12.7 Final Remarks

It is undeniable that the advent of the space age opened great prospects for the economic and social well-being of all human beings.[59] However, since the beginning of modern space activities, a new approach toward space-related matters appears to be necessary, mostly from a regulatory point of view. This new approach is intended to meet often conflicting needs of emerging space actors. Although "those who drafted the Outer Space Treaty intentionally kept its scope broad enough to govern all future space activities,"[60] it is equally important to point out all possible future needs that follow technological developments as the founding fathers of the Outer Space Treaty could not have envisaged all that the future could bring.

In particular, among those new needs, the paper mentions the international regulatory certainty necessary for important investments, the promotion of global public interest in outer space, and the maintenance of international peace and security. Besides such needs, we have also focused on some of the contemporary challenges to the most fundamental principles of space law.

More specifically, this paper identified possible actions toward the strengthening of the legal regime as a whole and, namely, both at the *stricto* sensu regulatory level (one comprehensive and modernized Law of Outer Space Convention) and at the organizational level (new Global Space Organization).

In the second part of the paper, we have focused on the model that should be followed to implement the idea of a comprehensive space law system and what institution should be the landmark in such new international regime in space law and policy.

[59] R. Jakhu, *Legal Issues*, cit., p. 4, who points out as "*The international law-making process has produced basic legal principles that represent a fair balance of interests between developed and developing countries. However, growing pressure by a number of countries for increased privatization, commercialization, deregulation, and globalization, along with recent changes in the global geopolitical situation, are creating disturbing disagreements about the interpretation of the Treaty, its implementation, and the direction of future legal development.*"

[60] R. Jakhu, *Legal Issues*, cit., p. 4.

Given that the UN (i.e., COPUOS) has always been the natural forum where space-related matters are being discussed and where international cooperation takes place, we uphold the idea that the UN should keep on playing a pivotal and irreplaceable role in regulating space activities. Moreover, taking into account that also nongovernmental and private actors start to play a crucial role in the field, the UNCLOS model could be considered as a viable example to enhance the current international space regime. As for the organizational side, the example of INTELSAT has been mentioned, even though it is limited to a specific sector of space activities.

MIX
Papier aus verantwortungsvollen Quellen
Paper from responsible sources
FSC® C105338
FSC
www.fsc.org

If you have any concerns about our products, you can contact us on
ProductSafety@springernature.com

In case Publisher is established outside the EU, the EU authorized representative is:
Springer Nature Customer Service Center GmbH
Europaplatz 3, 69115 Heidelberg, Germany

Printed by Libri Plureos GmbH
in Hamburg, Germany